S

Environmental health promotion

Sonja Kahlmeier

Environmental health promotion

development, implementation and evaluation

Südwestdeutscher Verlag für Hochschulschriften

Impressum/Imprint (nur für Deutschland/only for Germany)
Bibliografische Information der Deutschen Nationalbibliothek: Die Deutsche Nationalbibliothek verzeichnet diese Publikation in der Deutschen Nationalbibliografie; detaillierte bibliografische Daten sind im Internet über http://dnb.d-nb.de abrufbar.
Alle in diesem Buch genannten Marken und Produktnamen unterliegen warenzeichen-, marken- oder patentrechtlichem Schutz bzw. sind Warenzeichen oder eingetragene Warenzeichen der jeweiligen Inhaber. Die Wiedergabe von Marken, Produktnamen, Gebrauchsnamen, Handelsnamen, Warenbezeichnungen u.s.w. in diesem Werk berechtigt auch ohne besondere Kennzeichnung nicht zu der Annahme, dass solche Namen im Sinne der Warenzeichen- und Markenschutzgesetzgebung als frei zu betrachten wären und daher von jedermann benutzt werden dürften.

Verlag: Südwestdeutscher Verlag für Hochschulschriften GmbH & Co. KG
Heinrich-Böcking-Str. 6-8, 66121 Saarbrücken, Deutschland
Telefon +49 681 37 20 271-1, Telefax +49 681 37 20 271-0
Email: info@svh-verlag.de

Approved by: Basel, University, Diss., 2001

Herstellung in Deutschland:
Schaltungsdienst Lange o.H.G., Berlin
Books on Demand GmbH, Norderstedt
Reha GmbH, Saarbrücken
Amazon Distribution GmbH, Leipzig
ISBN: 978-3-8381-2991-4

Imprint (only for USA, GB)
Bibliographic information published by the Deutsche Nationalbibliothek: The Deutsche Nationalbibliothek lists this publication in the Deutsche Nationalbibliografie; detailed bibliographic data are available in the Internet at http://dnb.d-nb.de.
Any brand names and product names mentioned in this book are subject to trademark, brand or patent protection and are trademarks or registered trademarks of their respective holders. The use of brand names, product names, common names, trade names, product descriptions etc. even without a particular marking in this works is in no way to be construed to mean that such names may be regarded as unrestricted in respect of trademark and brand protection legislation and could thus be used by anyone.

Publisher: Südwestdeutscher Verlag für Hochschulschriften GmbH & Co. KG
Heinrich-Böcking-Str. 6-8, 66121 Saarbrücken, Germany
Phone +49 681 37 20 271-1, Fax +49 681 37 20 271-0
Email: info@svh-verlag.de

Printed in the U.S.A.
Printed in the U.K. by (see last page)
ISBN: 978-3-8381-2991-4

Content

Figure 2-2, p. 21: with friendly permission of the Evaluation Management and Resources Centre, Swiss Federal Office of Public Health

Figure 4-2, p. 51: with friendly permission of the Environment and Health Unit, Swiss Federal Office of Public Health (modified)

Zusammenfassung

Das Gebiet der „umweltbezogenen Gesundheit" („environmental health") behandelt diejenigen Aspekte der menschlichen Gesundheit und Krankheit, die durch Umweltfaktoren bestimmt werden. Das Gebiet umfasst nicht nur direkte Effekte von schädlichen Substanzen, sondern auch indirekte Auswirkungen der physischen und psychosozialen Umwelt auf Gesundheit und Wohlbefinden. Es beinhaltet auch die Beurteilung und Kontrolle von potentiell gesundheitsgefährdenden Umweltfaktoren. Die nationalen „Aktionspläne Umwelt und Gesundheit" (APUG), welche seit Mitte der 90er Jahre in ganz Europa entwickelt werden, sind ein neuartiger Versuch für integrierte Umwelt- und Gesundheitsprogramme. Der Schweizer APUG, welcher seit 1998 umgesetzt wird, war einer der ersten Aktionspläne für Umwelt und Gesundheit, der in einem industrialisierten Land entwickelt wurde. Er konzentriert sich auf die drei Themenbereich „Natur und Wohlbefinden", „Mobilität und Wohlbefinden" sowie „Wohnen und Wohlbefinden". Im Zusammenhang mit der Entwicklung, Umsetzung und Evaluation von solchen Programmen zur Förderung der umweltbezogenen Gesundheit („environmental health promotion programs") gibt es eine Reihe von offenen Fragen, mit denen sich diese Dissertation beschäftigt hat.

Im Zusammenhang mit umweltbezogener Gesundheit ist die Wohnqualität ein oft genanntes Thema. Die wissenschaftliche Basis für die Entwicklung von geeigneten Strategien zur Förderung von Wohnqualität und Wohlbefinden ist jedoch lückenhaft. Im ersten Teil dieser Dissertation wird eine Studie zu subjektiver Wohnqualität und Wohlbefinden präsentiert, die in der Nordwestschweiz durchgeführt wurde. Die Studie zeigte, dass eine höhere Zufriedenheit mit der Umweltqualität sowie mit der Wohnung selbst bei Umzügerinnen und Umzügern mit einem verbesserten Wohlbefinden assoziiert war. Die positive Assoziation mit Umweltindikatoren blieb auch bei denjenigen Teilnehmenden bestehen, die nicht wegen der Umweltqualität umgezogen waren. Es konnte jedoch nicht abschliessend geklärt werden, welcher Einzelfaktor der Umweltqualität dafür verantwortlich war: Die beiden Umweltindikatoren „Luftqualität" und „Lage des Hauses" schienen jeweils für eine Gruppe von verschiedenen Faktoren zu stehen. Daraus lässt sich schliessen, dass bei Projekten zur Förderung der Wohnqualität ein umfassender Ansatz angewendet werden sollte. Die jeweilige Ausgangslage und die Sicht der Betroffenen sollte dabei mit einbezogen werden.

Das Fehlen der wissenschaftlichen Basis ist jedoch nicht die einzige Schwierigkeit bei der Entwicklung von Programmen zur Förderung der umweltbezogenen Gesundheit. Eine allgemeine Diskussion von Stärken und Schwächen bei der Entwicklung und Umsetzung des Schweizer APUG hat gezeigt, dass seine Stärken in der Formulierung spezifischer Ziele in ausgewählten Themenbereichen, seinem Ansatz als eigentliches Förderungsprogramm für umweltbezogene Gesundheit und in der umfassenden Evaluation liegen. Die Förderung umweltbezogener Gesundheit ist immer eine intersektorielle Aktivität. Deshalb sollten idealerweise alle relevanten Akteure sowohl innerhalb als auch ausserhalb der Administration in die Entwicklung solcher Programm einbezogen werden, um die Zusammenarbeit sicher zu stellen. Es wurde gezeigt, dass während der Entwicklung des Schweizer APUG innerhalb der Administration eine gute Kollaboration erreicht wurde. Eine Schwäche der meisten APUG ist jedoch der mangelnde Einbezug der Bevölkerung und wirtschaftlicher Kreise sowie das Fehlen einer Umsetzungsstrategie mit angemessenen finanziellen Mitteln. Die grösste Herausforderung für diese prinzipiell wertvollen Programme liegt in der Sicherstellung der Verbindung zwischen Umwelt und Gesundheit auf struktureller Ebene über den intersektoriellen Entwicklungsprozess hinaus, um eine dauerhafte Allianz zu gewährleisten.

Evaluation sollte ein inhärenter Teil jedes Gesundheitsförderungsprogramms sein. Die umfassende Evaluation des Schweizer APUG besteht einerseits aus einer fortlaufenden Analyse des Umsetzungsprozesses (Prozessevaluation). Andererseits wurden basierend auf Wirkungsmodellen Indikatoren definiert, mit denen zielbezogene Resultate und einige indirektere Auswirkungen beurteilt werden (Outcome und Impact Evaluation). Eine 1999/2000 durchgeführte Erhebung der Ausgangslage zu diesen Indikatoren unterstrich den Handlungsbedarf in den drei Bereichen Mobilität, Wohnen und Natur. Aufgrund von Rückmeldungen aus der Prozessevaluation wurde 2001 ein Umsetzungsprogramm zum Schweizer APUG entwickelt. Während der Ausarbeitung dieses Umsetzungsprogramms wurde deutlich, dass die vorhandenen Ressourcen nicht ausreichen würden, um die formulierten Ziele für die drei Themenbereiche bis 2007 zu erreichen. Dementsprechend wurden die Ziele neu definiert, wobei man sich auf drei Pilotregionen beschränkte. Es wurde auch erkannt, dass eine langfristige Perspektive für das Erreichen einer wirklich intersektoriellen Zusammenarbeit und der strukturellen Veränderungen nötig sein wird.

Inzwischen begann die Weltgesundheitsorganisation (WHO) mit der Entwicklung eines Sets von Umwelt-Gesundheits-Indikatoren („environmental health indicators") für die internationale Anwendung. Als Beitrag zur Diskussion über verschiedene Vorgehensweisen bezüglich Umwelt-Gesundheits-Indikatoren und deren Anwendungen wurde das WHO Indikatorenset mit den Indikatoren für die Evaluation des Schweizer APUG verglichen. Ausserdem wurde die Eignung eines internationalen Indikatorensets für die Evaluation nationaler Programme diskutiert. Das von der WHO vorgeschlagene Umwelt-Gesundheits-Indikatorenset dient einer strukturierten Darstellung der Ursachen-Wirkungsketten. Das Set ist nützlich für das Monitoring und internationale Vergleiche der allgemeinen Umwelt- und Gesundheitssituation und unterstützt deshalb die Prioritätensetzung. Eine Reihe methodischer und technischer Schwierigkeiten muss jedoch beachtet werden, insbesondere bezüglich einer Abschätzung von gesundheitlichen Auswirkungen. Die Indikatoren für die Evaluation des Schweizer APUG wurden von bereits formulierten Programmzielen abgeleitet, während Umwelt-Gesundheits-Indikatoren im Gegensatz dazu zur Prioritätensetzung und Zielformulierung führen sollen. Die Relevanz international entwickelter Indikatoren ist ausserdem je nach nationalem Kontext unterschiedlich; und sie erlauben auch keine Evaluation des Umsetzungsprozesses. Umwelt-Gesundheits-Indikatoren sind deshalb für die Evaluation nationaler Programme nur beschränkt geeignet.

Für die Zukunft liegt die Herausforderung in der Ausarbeitung eines Umwelt-Gesundheits-Indikatorensets, welches sowohl internationale Vergleiche erlaubt als auch den nationalen Prioritäten entspricht, sowie in der Entwicklung von Gesundheits-indikatoren im Rahmen des Monitoring der nachhaltigen Entwicklung in industrialisierten Ländern wie der Schweiz.

Summary

Environmental health deals with those aspects of human health and disease that are determined by factors in the environment. It does not only include direct effects of harmful substances but also more indirect consequences of the physical and psychosocial environment on health and wellbeing. It also comprises the assessment and control of environmental factors which can potentially affect health. The "National Environment and Health Action Plans" (NEHAPs), which have been developed throughout Europe since the middle of the 1990s, are a novel attempt for an integrated environment and health policy. The Swiss NEHAP, which is implemented since 1998, was among the first to be developed in an industrialized country. It focuses on the three topic "Nature and Wellbeing", "Mobility and Wellbeing" and "Housing and Wellbeing". There are a number of open issues in relation to the development, implementation and evaluation of such environmental health promotion programs, which were addressed in this thesis.

Housing quality is often named as a key area in environmental health. However, the scientific basis for the development of appropriate promotion strategies on housing quality and wellbeing is incomplete. In the first part of this thesis, data from a study on perceived housing quality and wellbeing, which was carried out in the north-western Region of Switzerland, is presented. The study showed that a higher satisfaction with environmental housing quality and with the apartment was associated with an improved wellbeing of movers. The positive association with environmental indicators was persistent in participants who had moved for other than environmental reasons. However, it could not be entirely clarified which single factors in the residential environment were most influential. Both environmental indicators "perceived air quality" and "location of the building" seemed to reflect a group of different determinants. It can be concluded that an integrated approach should be applied in projects aiming at the improvement of the housing quality, taking the respective situation and views of the ones affected into account.

But the lack of scientific evidence is not the only challenge in the development of environmental health promotion programs. A general discussion of strengths and weaknesses of the development and implementation process of the Swiss NEHAP showed that the strengths of the Swiss NEHAP lie in the formulation of specific targets in selected areas, its approach as a environmental health promotion program, and its comprehensive evaluation. Environmental health promotion is always an intersectorial activity. Therefore, all relevant actors, ideally within as well as outside the administration, should be involved into the development of such programs to ensure their collaboration. It was shown that a good inter-administrational involvement was achieved in the development process of the Swiss NEHAP. Weaknesses in most NEHAPs are the lack of involvement of the general public and of the economic sector, and the absence of an implementation strategy along with adequate financing. The greatest challenge in the implementation of this in principal valuable framework will be to ensure the link between health and environment on a structural level beyond an intersectorial development phase to build a real and long-term stable alliance.

Evaluation should be an inherent part of every health promotion program. The comprehensive evaluation of the Swiss NEHAP consists on the one hand of the continuous analysis of the implementation of the program (process evaluation). On the other hand, indicators were defined based on impact models to assess aim-related outcomes and a selected number of more distal impacts (outcome and impact evaluation). The baseline assessment of these indicators in 1999/2000 underlined the need for action in the three topics Mobility, Housing, and Nature. As a major consequence of feedback from the process evaluation, an implementation program for the Swiss NEHAP was developed in 2001. During the development of this implementation program, it became apparent that it would not be possible to reach the aims formulated for the three topics until 2007 on a national level with the resources at hand. Consequently, the objectives were redefined focusing on three pilot regions. It has also been recognized that a long term perspective will be necessary to achieve truly intersectorial collaboration and structural changes.

Meanwhile, the World Health Organization (WHO) started with the development of a set of environmental health indicators for international application. As a contribution to the ongoing discussion on the different approaches in relation to environmental health indicators and their application, the WHO indicator set was compared with the Swiss evaluation indicators. Additionally, the suitability of an international indicator set for the evaluation of national programs was discussed. The set of environmental health indicators (EHIs) proposed by the WHO serves a structured description of the underlying cause-effect chains. The set is useful for monitoring and international comparison of the general environment and health situation, thus supporting priority setting. However, a number of methodological and technical difficulties need to be addressed, particularly in relation to health impact assessment. Indicators for the evaluation of NEHAPs were derived from previously formulated policy targets while EHIs, in contrast, should lead to priority setting and policy formulation. Additionally, the relevance of internationally developed indicators will vary in the national context and they do not allow to evaluate the policy implementation process. Therefore, the suitability of EHIs for the evaluation of national environmental health promotion programs is limited.

Challenges for the future lie in the development of a set of environmental health indicators, which allows international comparisons and at the same time responds to national priorities, and in the elaboration of health indicators in the framework of sustainable development monitoring in industrialized countries such as Switzerland.

PART I:

INTRODUCTION AND BACKGROUND

1   Introduction

1.1   Environmental health promotion: open issues

Environmental health has been defined as "those aspects of human health and disease that are determined by factors in the environment".[1] It also includes the assessment and control of environmental factors which can potentially affect health. It has been estimated that 25 to 33% of the global burden of disease can be attributed to environmental risk factors.[2] Even taking into account the considerable uncertainties immanent in such estimates, the percentage might be too low since it includes only the proportion of disease and not the total proportion of ill health. But environmental health does not only comprise direct effects of e.g. chemicals, radiation or accidents but also more indirect effects of the physical, psychological and social environment on health and wellbeing, comprising a large variety of determinants, such as urban development, land use, transport, or housing.[1]

Uncertainties are frequent in environmental health estimates since precise measures of the underlying cause effect relationships are still rare.[3, 4] One example is housing quality which is often named as a key area in environmental health in developing as well as in developed countries.[5-8] The association between physical determinants of housing quality such as crowding, dampness or the access to piped water and indicators of disease such as asthma or diarrhoeal diseases have been well established.[5, 9] However, a more comprehensive concept to analyse the various dimensions of the construct "housing quality" is lacking.[10-12] Additionally, only few studies investigated the association between different aspects of housing quality and non-disease related dimensions of health such as wellbeing, which are likely to be more affected in developed countries.[13] Therefore, the scientific basis for the development of appropriate promotion strategies e.g. on housing quality and wellbeing, is often incomplete.

Due to the large variety of factors influencing health and wellbeing, they should not be on the agenda of the health sector alone, but an intersectorial approach is needed in the promotion of environmental health.[14] The "National Environment and Health Action

Plans" (NEHAPs), which have been developed throughout Europe since the middle of the 1990s, are a novel attempt for an integrated environment and health policy.[7] In practice, however, the implementation of such environmental health promotion programs is challenging. Competences and finances are usually allocated to specific topics within the various ministries, thus complicating joint action. Intersectorial administrative structures to address such problems in an integrated way are often missing and cooperation across administrative boundaries is not yet the rule.[15, 16]

According to the "Public Health Action Cycle",[17] evaluation should be central in every health promotion program. Ideally, the evaluation should induce a learning process to improve current activities and enable better planning of future action.[18, 19] Being already a challenging task in classical health promotion,[20-22] in environmental health promotion evaluation is confronted with additional difficulties such as uncertainties on cause effect chains, lack of adequate data or complex program implementation structures.

1.2 Objectives and content of this thesis

This thesis deals with open issues in the field of environmental health promotion. More specifically, the following research questions will be addressed:

1. How can associations between different determinants of housing quality and wellbeing be measured?
2. Which dimensions of housing quality are associated with wellbeing in an industrialized country like Switzerland?
3. How important is the perceived environmental housing quality which could be addressed by an environmental health promotion program?
4. How can the explicit linking of health promotion and environmental protection be translated into an environmental health promotion program?
5. How can such an environmental health promotion program be evaluated?

In the first part of this thesis, data from a study on perceived housing quality and wellbeing, which was carried out in the north-western Region of Switzerland, is presented and discussed. This study provides insight into a field of environment and health where detailed information is scarce (*research questions 1 to 3*). Subsequently, requirements and problems in the development and implementation of environmental health promotion programs are discussed in general, exemplified by the Swiss NEHAP, and first conclusions are drawn (*research question 4*). The following part of the thesis describes the evaluation of the Swiss NEHAP and discusses the suitability of environmental health indicators for policy evaluation (*research question 5*). Finally, the main findings of this thesis are summarised and the implications for future activities are discussed.

2   Background

In the following, the relevant theoretical background and the key concepts used later on in this thesis are introduced. In the first paragraph milestones in development of the field "environmental health" are outlined and key concepts are described. Subsequently, a short introduction in evaluation theory is given.

2.1   Development and key concepts of environmental health promotion

2.1.1   Health and health promotion

In 1948, health had been defined by the World Health Organisation (WHO) as "a state of complete physical, mental and social wellbeing and not merely the absence of disease or infirmity".[23] In the last two decades however, this static definition has developed into a more dynamic concept with no clear-cut dividing line between health and disease. The positive point of view has been underlined by focusing on the prerequisites of health rather than the risk factors of disease: "Health is a positive concept emphasizing social and personal resources, as well as physical capacities. (…) Political, economic, social, cultural, environmental, behavioural and biological factors can all favour health or be harmful to it."[14]

At the First International WHO Conference on Health Promotion in Ottawa, health promotion has been defined in the "Ottawa Charter" as "the process of enabling people to increase control over, and to improve, their health".[14] Furthermore, the following principles should be applied:

- interdisciplinary cooperation of all sectors within and outside the health care system by putting health care on the agenda of all sectors and at all levels,
- coordinated action by all concerned (individuals, communities, institutions, administration, politics, economic sectors and industry, nongovernmental organisations (NGOs), media etc.),

- participation of the ones affected in planning, development and implementation of projects,
- empowerment by strengthening of self-confidence and the ability to cope with problems to increase options to exercise control over ones health and environment and to make choices conducive to health.

Health promotion programs should aim at influencing individual behaviours as well as the political, organisational, social and environmental conditions to facilitate "healthy choices".[14]

It had already been mentioned in the Ottawa charter in 1986 that "the protection of the natural and built environments and the conservation of natural resources must be addressed in any health promotion strategy".[14] Ten years ago, however, a special emphasis was laid on the need to create supportive environments for health at the Third International Conference on Health Promotion. One of the key public health action strategies named at this conference was to "build alliances for health and supportive environments in order to strengthen the cooperation between health and environment campaigns and strategies".[24] In the Sundsvall-statement endorsed at this conference, education, transport, housing and urban development, industrial production and agriculture were identified as priorities for action.

2.1.2 Environmental health promotion

Intuitively, the association between the environment and the health of individuals had long been known.[25] One of the first environmental epidemiology studies was published in 1767 on serious health consequences of the consumption of cider which had been contaminated with lead during the production process.[26] A formal recognition of the relationship between environment and health, however, followed almost 100 years later after a cholera outbreak in London which could be related to the drinking water provided by one specific waterworks.[27] Epidemiology as a scientific discipline, using systemized principles in the design and analyses of studies, evolved only in the second half of the twentieth century.[28, 29]

Accordingly, the promotion of environmental health in an integrated way also developed only recently.[30] Early concepts emerged in the 1960s,[31] but the starting point for the development of promotion programs dealing with environmental pollution and health consequences in Europe was the WHO "Health for All" strategy launched in 1982.[32] This strategy laid the basis for a European health policy and common aims for the year 2000. In the updated version of the strategy, nine of the 38 aims related to environment and health.[6] The concept was specified further in 1989 at the first European Conference on Environment and Health, where it was stated that environmental health included both direct pathological effects of chemicals or biological agents as well as (often indirect) effects of factors like housing, urban development and transport.[1] A more comprehensive and political perspective was introduced by the concept of sustainability, incorporating economic development, environmental protection and social justice.[5, 33] Each of these three dimensions of sustainability can have an impact of human health and wellbeing as shown in Figure 2-1.

Figure 2-1:   *Impact of the three dimensions of sustainability on health and wellbeing.*
                 (translated from[34])

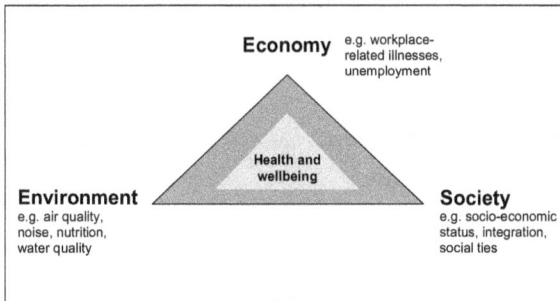

The association between environment and health was one of the key topics at the United Nations Conference on Environment and Development in 1992 in Rio de Janeiro. In preparation of this conference, the WHO created the Commission for Health and Environment. Its report "Our planet, our health"[5] contributed significantly to the formulation of environmental health promotion measures in the "Agenda 21", the "action plan for a sustainable development in the 21st century" adopted at the conference.[33] In chapter six of Agenda 21, the protection and promotion of human health was specifically addressed, stating that "human health depends on a healthy environment, clean air and clean water, waste disposal and proper nutrition". The WHO was assigned as leading organisation for the implementation of this chapter of the Agenda 21. Subsequently, the European WHO member states were appealed to develop their own National Environment and Health Action Plans (NEHAPs) based on the European Action Plan Environment and Health adopted in 1994[7] and as part of the practical implementation of sustainable development.

2.2   Introduction in evaluation theory

Evaluations are carried out in various fields today and the term is defined quite broadly. Policy evaluation, however, implies the analysis of the efficacy of an intervention. In a very general sense, evaluations can be classified into the following three **types**:[35]

- needs assessment to identify problems or goals which can include the conceptualisation and design of an intervention,
- monitoring of the development or implementation of a program (also called "process" or "formative" evaluation)
- prospective or retrospective assessment of usefulness and effects of a program (also called "summative" evaluation).

Assessments that comprise formative as well as summative questions are also known as "comprehensive" evaluations.

Evaluations can be classified as well according to their role in a project lifecycle:

Figure 2-2:   The role of evaluation in the lifecycle of a project.[36]

Agenda-setting

Efficiency   Relevance

Accountability ← EVALUATION → **Formulation**

Effectiveness   Process

Implementation

In this classification, evaluation is seen as a cyclic process. First, the relevance of the problem in question is assessed. Then, the implementation process is analysed, which can include an assessment of the appropriateness of the intervention. The effectiveness of the project in reaching its aims is assessed on the basis of the compliance between intended and actual conditions or behaviours. Finally, it can be analysed whether the costs of a project were adequate in relation to its benefits (efficiency). Based on the evaluation results, a new intervention can be planned.[36] In the evaluation of environmental health promotion programs, however, a cost-benefit analysis requires large efforts due to a frequent lack of adequate data and uncertainties e.g. in valuating a life year lost or in the quantification of intangible costs, e.g. pain, suffering.[37]

Different policy elements can be of interest in an evaluation:[19, 35, 36] While during the agenda-setting and formulation of a program, the policy concept and the intervention design are analysed, administrative arrangements are important elements for the process evaluation. Outputs are all physical, informal or service products of a program. Effects of an invention can be classified into different elements. In this thesis, the term "outcome" will be used for directly aim-related changes in behaviours or conditions, while the totality of - intentional or unintentional - effects, including also more distal changes, are named "impacts".

In most cases, a goal-oriented approach is part of the evaluation. It is therefore important to distinguish between general goals which are often vaguely formulated ("reduction", "improvement" without further specification) and do not allow a methodological evaluation, and operationalized objectives or targets which specify the desired results in terms of time and magnitude. Those types of aims are often referred to as "SMART objectives", which means that they should be:[36]

- specific,
- measurable,
- appropriate,
- realistic, and
- time bound.

Useful tools for goal-oriented evaluations are "impact models".[35] They consist of a number of hypotheses on the expected relationship between a program and its objectives and serve as a basis for an understanding why measures reached their objectives or what eventually hindered their effect. They contain:

- a causal hypothesis, which describes the influence of various determinants on behaviours or conditions that the intervention seeks to modify,
- an intervention hypothesis, which specifies the expected relationship between intervention and determinants mentioned in the causal hypothesis and
- an action hypothesis, which explains why a change in the mentioned determinants is believed to lead to a change in the behaviours or conditions. This last step facilitates the inclusion of influence factors which have not been comprised in the program but might affect the attainment of the objectives.

The following hypothetical impact model illustrates the approach with the example of a bicycle promotion program for commuters.

*Table 2-1:   Hypothetical impact model for a bicycle promotion program for
                 commuters.[38]*

causal hypothesis and determinants	The share of commuters using the bicycle to go to work is highest in companies which make bicycle use most attractive by offering various incentives.
	possible incentives (determinants): e.g. number, location and quality of bicycle stands, existence of changing rooms and showers, reimbursement of kilometres driven to work by bicycle, restrictive handling of car parking spaces etc.
intervention hypothesis	Incentives such as the installation/renovation of bicycle stands, the provision of changing rooms etc. can increase the attractiveness of the bicycle use in comparison with the use of a car and therefore can lead to an increased use of the bicycle to go to work.
action hypothesis and additional influence factors	Incentives will lead to an increased use of the bicycle because it has not been attractive enough to use it so far.
	↳additional influence factors on the bicycle use: e.g. distance from home to the company, security of roads which need to be used, availability of public transport, image of bicycle in the company etc.

When the impact model has been formulated, the design of the evaluation can be specified. Three approaches can be differentiated:[39]

- descriptive (How has a project developed? Which projects have been carried out?)
- normative (In how many percent of projects a certain standard has been reached? How many people were reached by a project? Have the program objectives been attained?)
- causal (To what extent the project has contributed to the attainment of objectives?)

Each approach implies a different evaluation strategy and different research methods:[19] While a descriptive approach is based on qualitative research techniques and comprises descriptions of the relevant issues, a normative approach implies a distance-to-target-comparison typically based on statistical information. A causal approach is based on e.g. control-group-studies, cross sectional or longitudinal studies. In policy evaluation, however, a causal approach can rarely be applied since numerous influence factors that are not under the program's control usually do not allow a clear allocation of effects. In health promotion evaluation, this problem is referred to as the "control group dilemma":[20-22] Health promotion – and especially environmental health promotion – is often carried out in settings like a city neighbourhood or even a region, which are not closed systems but open to external factors which can interfere with an intervention. Randomised assignment to an intervention and a control group is often impossible.

Additionally, such settings are open to everyone and subjects from a "control" group can have access to activities as well or read about it in the media, which leads to a "contamination" of the control group.

Based on the impact model and according to the chosen approach, the identification of indicators is the next step in the preparation of an evaluation concept. Indicators should be valid (i.e. they should measure what they are supposed to measure), reliable (i.e. the results should be reproducible), sensitive to changes and as specific as possible to changes in the situation concerned.[18] An essential step in the evaluation of comprehensive programs like the NEHAPs is a review of available data to increase the efficiency in gathering the necessary data and to benefit of available knowledge and experience from existing studies.

PART II:

DEVELOPMENT AND IMPLEMENTATION OF

ENVIRONMENTAL HEALTH PROMOTION

PROGRAMS

Introduction

In the Swiss NEHAP, the following three main topics were selected (see also chapter 4.2):[40]

- "Nature and Wellbeing", dealing with agriculture and nutrition,
- "Mobility and Wellbeing, and
- "Housing and Wellbeing".

During the development and formulation of goals, objectives and measures for each topic, it became apparent that especially in the field "Housing quality and Wellbeing", the theoretical basis was relatively weak. At the same time, a study on housing quality and the reasons for small scale migration was carried out in the north-western region of Switzerland. It was possible to include a few questions on the wellbeing of the subjects into this study and to explore this topic along with a detailed set of housing quality indicators. The results of this study, with a special focus on perceived environmental housing quality and wellbeing, are presented in the first section of part II (chapter 3).

A general discussion of strengths and weaknesses of the development and implementation process of the Swiss NEHAP and first lessons for environmental health promotion programs are presented in the second section of this part in chapter 4.

3   Perceived environmental housing quality and wellbeing of movers[*]

Abstract

Study objective: To examine whether changes in environmental housing quality influence the wellbeing of movers taking into account other dimensions of housing quality and sociodemographic factors.

Design and setting: Cross sectional telephone survey (random sample of 3870 subjects aged 18-70 who had moved in 1997, participation rate 55.7%.) in the north-western region of Switzerland, including the city of Basel. Associations between changes in satisfaction with 40 housing quality indicators (including environmental quality) and an improvement in self rated health (based on a standardized question) were analysed by multiple logistic regression adjusting for sociodemographic variables. Objective measures of wellbeing or environmental quality were not available.

Results: A gain in self rated health was most strongly predicted by an improved satisfaction with indicators related to the environmental housing quality measured as "location of building" (adjusted odds ratio (OR) =1.58, 95 % confidence interval (CI) =1.28-1.96) and "perceived air quality" (OR=1.58, 95% CI=1.24-2.01) and to the apartment itself, namely "suitability" (OR=1.77, 95% CI=1.41-2.23), "relationship with neighbours" (OR=1.46, 95% CI=1.19-1.80) and "noise from neighbours" (OR=1.32, 95% CI=1.07-1.64). The destination of moving and the main reason to move modified some of the associations with environmental indicators.

Conclusion: An improvement in perceived environmental housing quality was conducive to an increase in wellbeing of movers when other dimensions of housing quality and potential confounders were taken into account.

[*]   *Published as: Kahlmeier S, Schindler C, Grize L, Braun-Fahrländer C: Perceived environmental housing quality and wellbeing of movers. J Epidemiol Community Health 2001;55:708-715.*

3.1   Introduction

In many cities in developing countries, inadequate housing, lack of sanitation, dampness or overcrowding endanger the health of inhabitants, especially among economically disadvantaged groups.[5, 9] In industrialized countries too, relations between housing quality and health were reported. A large body of research focused on specific aspects of housing quality like e.g. dampness and specific health outcomes such as respiratory health.[41, 42] Others applied a broader concept of housing quality and/or more general concepts of health. E.g. Haan et al. demonstrated that residence in a poor neighbourhood was associated with an approximately 50% increase in mortality compared to a non-poverty area.[43] Yen and Kaplan showed that living in low social environments was associated with both, an increased risk of death[12] and decreased self rated health.[11] They also reported an increase in depressive symptoms. Malmström et al. found an association between neighbourhood socioeconomic environment and self rated health as well.[10] Mackenbach et al. showed that the presence or absence of housing problems was associated with both ill and excellent health, respectively.[44] A body of research focused on the impact of housing quality on health and wellbeing among the elderly, showing associations with mortality,[45] with different measures of wellbeing,[46] with life satisfaction and happiness,[47] and with self rated health.[45, 48] In many of these studies, self rated health has served as a useful summary measure of general wellbeing: It is associated with morbidity[49, 50] and mortality,[51] as well as with the use of physician services,[52] and with mental health.[53] In addition, self rated health also reflects aspects of social role, self-image,[54] and perceived control.[55]

Due to the growing body of evidence relating housing quality to wellbeing and health, the issue has been politically recognized in industrialized countries too.[8, 13, 56] This resulted in initiatives like the Healthy Cities Project, which was developed in 1986 to apply the Health for All principles at the local level in urban settings.[56] One of the qualities a Healthy City should aim to provide is a high quality physical environment, including housing quality.

Most of the studies on housing quality, health and wellbeing focused either on very specific single aspects such as dampness and asthma, not allowing conclusions on the overall impact of housing quality on general wellbeing or on proxy measures (like "poverty") or summary indicators of housing quality (like "presence or absence of housing problems in general"). But the question arises as to which of the different aspects of the complex construct "housing quality" are influential for the general wellbeing of citizens.[10, 11] The environmental quality of the housing surroundings may be an important component and in the public debate, environmental housing quality is often cited as the main driving force for suburbanisation processes.[57] Within a detailed set of indicators for different dimensions of housing quality, we therefore focused on indicators of environmental housing quality such as perceived noise and perceived air quality. Applying a more general concept of health, we studied if changes in these indicators were predictive of changes in self rated health as measure of wellbeing among movers in Switzerland after adjusting for changes in other indicators of housing quality (e.g. relating to the apartment itself or to infrastructure) and potential sociodemographic confounders.

3.2   Methods

The study was carried out in the north-western region of Switzerland including the city of Basel with approximately 200'000 inhabitants. The north-western region of Switzerland encompasses an area extending approximately 30 kilometres east and south of Basel with roughly 345'000 inhabitants. In summer 1998, a random sample of 3870 non-institutionalised adults, aged 18 to 70 years, with Swiss citizenship or permanent residence permit who had moved once in 1997 either within the city of Basel or out of the city of Basel into the north-western region of Switzerland was drawn from the population registry. Since this registry contains complete information on address changes, eligible persons could be traced. The random sample, stratified by type of mover (within the city vs. out of the city), was drawn in two stages: first, households were selected and second, the interview partner within each household was determined. Specially trained interviewers performed the standardized telephone interviews in August and September 1998. For 653 persons (16.9%) no valid phone number was available, 374 persons (9.6%) declined to participate, 282 persons (7.3%) did not live at the recorded address anymore, 223 persons (5.8%) could not be contacted during the whole interview period within up to 20 attempts, and 181 persons (4.7%) could not be interviewed due to other reasons (i.e. language). Information was thus obtained from a total of 2157 subjects (55.7%).

The questionnaire was based on existing questionnaires,[58-60] and pretested in a smaller sample. The study was introduced to the participants as a survey on the reasons for moving, the issues presented here were not mentioned. Demographic and socioeconomic information as a potential source of bias was collected on sex, age, household composition, monthly household income, education, and type of moving (details see table 3-1). Next, participants were asked an open question about the main reason for moving. Answers were noted literally and then, according to prescribed rules, assigned to five main categories: (1) "apartment" (e.g. too small / big / expensive), (2) "personal reasons" (e.g. aging, marriage), (3) "neighbourhood" (e.g. not suitable for children, problems with neighbours or owner of the house, dirt, no parking space), (4) "environment" (e.g. perceived noise or air quality, traffic, not enough green) and (5) "political or social reasons" (e.g. school quality, taxes). This question was answered by 2000 participants. For the analyses, the reasons to move were dichotomised into "environmental reasons" (categories 3 and 4) and "other reasons" (categories 1, 2 and 5). Participants were then asked about their present self rated health and the change in self rated health was assessed with the question: "And how is that in comparison to your former residence. Do you now feel in general better, about the same or worse?". Furthermore, they had to assess 40 indicators of housing quality both for their former and their present residence. Besides the environmental quality, these indicators regarded the apartment itself, infrastructure and community services as well as educational and leisure time opportunities. A complete list of all indicators is given in figure 1. The Swiss school grading scale being familiar to everyone living in Switzerland, with grades from 1 to 6, was used for the assessment (1=very bad, 6=very good, 4=sufficient, half grades were allowed).

3.2.1 Analyses

The change in self rated health was used as outcome measure. It was dichotomised into the categories "improved" and "not improved" (the latter including "no change" and the small group reporting a deterioration). For each of the 40 housing quality indicators the difference between the actual and the former residence was calculated and likewise dichotomised into "improved" and "not improved". Out of the 2157 respondents, 13 had missing values in the outcome variable. For 1961 subjects we had complete information on outcome and all sociodemographic variables. However, answers were missing on some of the housing quality indicators. But for none of the 40 indicators, subjects with missing information differed significantly from those with no improvement as far as changes in self rated health were concerned. Therefore, missing values were coded as "not improved" in order not to reduce the sample size further. The multivariate analyses were thus based on a total of 1961 subjects.

Descriptive analyses
The data were first analysed by means of cross tabulations of the change in self rated health (improved/not improved) by sociodemographic variables and by the differences in the housing quality indicators (improved/not improved). The degree of heterogeneity across subgroups was evaluated with the Chi-square-test and the odds ratios for the cross-tabulations were estimated using logistic regression.

Dimensions of housing quality
Next, we performed a factor analysis (varimax rotation).[61] The indicators could be grouped into 8 dimensions of housing quality (factors).To study the relative importance of these different housing quality dimensions as potential determinants of the change in self rated health (dependent variable), a logistic regression analysis was performed, including the standardized factor scores along with the sociodemographic covariates sex, age, household composition, household income, education, and type of moving.

Logistic regression of individual housing quality indicators
Subsequently, we evaluated which of the 40 single indicators were most influential for a change in self rated health. Starting from a logistic regression model including the sociodemographic covariates and all 40 housing quality indicators, we eliminated indicators with p-values >0.20. This resulted in a final model with 14 indicators (question verbatim see annex).

Logistic regression in subgroups

To investigate whether associations between changes in self rated health and changes in "environmental" housing quality indicators were different between those who moved within the city as compared to those who moved out of the city or between those who moved for "environmental reasons" compared to those who moved for "other reasons", we ran stratified logistic regression analyses. Effect modification was evaluated with the Chi-square-test for heterogeneity of estimates. With the same approach, we also studied whether moving from a multiple dwelling into a single family home, or owning the house or apartment since having moved modified the associations. The statistical software package SYSTAT 7.0[62] was used to perform the analyses.

Table 3-1: Sociodemographic characteristics of the sample, association with an improved self rated health (SRH) since having moved and frequency of environmental reasons as main reason to move. n=2144

	total		improved SRH since having moved				environmental reason *		
	number	%	number	%	OR	95% CI	number†	%	p‡
Total	2144	100.0	1230	57.4			428	21.4	
Sex									
Men	1022	52.3	555	54.3	1.00		208	21.8	
Women	1122	47.7	675	60.2	1.27	1.07-1.51	220	21.1	0.709
Age									
18-30 years	796	37.1	439	55.2	1.00		120	16.3	
31-45 years	930	43.4	528	56.8	1.07	0.88-1.29	209	23.8	
46-60 years	321	15.0	204	63.6	1.42	1.09-1.85	78	26.3	
61-70 years	97	4.5	59	60.8	1.26	0.82-1.94	21	24.1	<0.001
Household composition									
single adult	700	32.7	388	55.4	1.00		134	20.7	
2+ adults without children	901	42.0	501	55.6	1.01	0.83-1.23	158	18.7	
2+ adults with children	481	22.4	300	62.4	1.33	1.05-1.69	120	26.5	
single adult with children	62	2.9	41	66.1	1.57	0.91-2.71	16	28.6	0.006
Household income									
< 3000 SFr.	189	8.8	111	58.7	1.00		36	21.2	
3000 to 4999 SFr.	472	22.0	269	57.0	1.22	0.86-1.74	93	21.6	
5000 to 7499 SFr.	564	26.3	331	58.7	1.14	0.87-1.49	126	24.0	
7500 to 9999 SFr.	341	15.9	199	58.4	1.22	0.94-1.58	62	18.7	
≥ 10000 SFr.	398	18.6	214	53.8	1.21	0.90-1.61	74	19.6	0.370
missing	180	8.4							
Education									
high	891	41.6	475	53.3	1.00		153	18.7	
middle	1074	50.1	650	60.5	1.34	1.12-1.61	236	23.4	
low	164	7.6	96	58.5	1.24	0.88-1.73	37	23.4	0.040
missing	16	0.7							
Type of moving									
within the city	1011	47.2	539	53.3	1.00		174	18.8	
out of city	1133	52.8	691	61.0	1.37	1.15-1.63	254	23.7	0.008

* Compared to "other reasons"; includes the categories "environment" (for example, noise, traffic, not enough green) and "neighbourhood" (for example, suitability for children, problems with neighbours, dirt). † Based on a total of 2000 answers on the main reason to move ‡ χ^2 test

33

3.3   Results

The majority of the subjects (1230 of the 2144 participants, 57.4%) stated that in general their self rated health had improved compared to their former residence. 829 subjects (38.7%) reported no change and only a proportion of 3.9% (85 participants) reported a deterioration. An overview of the sociodemographic characteristics of the sample and of the associations with an improved self rated health since having moved is given in table 3-1).

Subjects who had moved out of the city, women, respondents with a middle education, respondents (two or more) with children and the 46 to 60 year olds were more likely to state that, in general, their self rated health had improved since they had moved. As also shown in table 3-1, some differences across sociodemographic subgroups were also found regarding the main reason to move: "environmental reasons" were mentioned more often by persons having moved out of the city, participants with children, participants with low or middle education, and in the age groups over 30.

3.3.1   Dimensions of housing quality and improved self rated health

A factor analysis was performed to study groupings of the 40 indicators. Figure 1 shows the 8 dimensions of housing quality having been identified. The label assigned to each factor intends to describe the respective dimension (figure 3-1, in quotation marks). The presented model explained 48.7% of the total variance in the 40 indicators. Improved self rated health was most strongly associated with an improved satisfaction with the two dimensions directly relating to the dwelling, namely the dimension "Apartment or building" (adjusted OR: 1.55, 95% CI 1.40-1.71) and the "Apartment-related social components" (1.52, 1.37-1.67), followed by an improved assessment of the dimension "Environment" (1.47, 1.33-1.63) and aspects relating to "Leisure time" (1.39, 1.25-1.55). An increased satisfaction with the dimensions "Suitability for children" (1.24, 1.04-1.48), "Community services" (1.17, 1.06-1.29), "Infrastructure" (1.16, 1.04-1.28) and "Cultural and social life" (1.12, 1.01-1.24), respectively, showed weaker associations with an improvement in self rated health since having moved.

Figure 3-1: Result of the factor analysis: 8 dimensions of housing quality with the corresponding variables and factor loads (in parentheses). n=2157

"Suitability for children"
- Suitability of surroundings for children (0.85)
- Suitability of surroundings for teenagers (0.80)
- institutionalised day-care (0.62)
- private day-care (0.56)
- school/kindergarten (0.75)
- availability of playgrounds (0.79)
- way to school (0.50)

"Apartment or building"
- comfort of the apartment (0.77)
- suitability of the apartment (0.61)
- condition of the apartment (0.83)
- condition of the building (0.77)

"Environment"
- perceived air quality (0.61)
- perceived traffic noise (0.76)
- location of the building (0.50)
- negative effects of traffic (0.75)
- perceived noise from airplanes (0.41)

"Cultural and social life"
- cultural life (0.63)
- possibilities to go out (0.78)
- organized home care (0.41)
- possibilities for adult education (0.60)
- clubs/associations (0.49)
- meeting places/community centres (0.51)

"Leisure time"
- equipment with parks / free spaces (0.61)
- „green" neighbourhood (0.58)
- sports facilities (0.68)
- security of surroundings (0.40)
- parking spaces (0.56)
- supply / security of bicycle lanes (0.55)
- supply / security of pavements (0.40)

"Community services"
- waste removal (0.65)
- maintenance of streets (0.65)
- cleanliness of surroundings (0.44)

"Infrastructure"
- facilities for daily shopping needs (0.73)
- postal offices / banks (0.72)
- medical supply (0.61)
- supply with public transport (0.65)
- way to work (0.41)

"Apartment-related social components"
- rent / mortgage (0.61)
- relationship with neighbours (0.63)
- noise from neighbours (0.46)

3.3.2 Individual housing quality indicators and improved self rated health

Table 3-2 shows the odds ratios for an improved self rated health associated with a higher satisfaction with the remaining 14 single indicators (out of the originally 40, see annex) since having moved. The indicators are grouped according to the results of the factor analyses (see table 3-2).

In the multivariate analyses, all associations were weaker than in the bivariate analyses and some associations even became borderline or non-significant. Nevertheless, 5 indicators remained significantly associated with an improved self rated health: In addition to the two "environmental" indicators "location of the building" and "perceived air quality" these included "suitability of the apartment", "relationship with neighbours" and "perceived noise from neighbours".

Table 3-2:   Association between an improved satisfaction with housing quality
indicators at the new residence and an improvement in self rated health
(SRH) since having moved

Improved satisfaction with*:	total n=2144		Improved SRH since having moved n=1961			
			unadjusted		adjusted[‡]	
	nr.	%	OR	95% CI	OR	95% CI
"Environment"[†]						
location of the building	1292	60.3	2.64	2.21-3.15	1.58	1.28-1.96
perceived air quality	1237	57.7	2.38	2.00-2.84	1.58	1.24-2.01
"Apartment or building"[†]						
suitability of the apartment	1352	63.1	2.81	2.35-3.37	1.77	1.41-2.23
comfort of the apartment	1414	66.0	2.18	1.82-2.61	1.26	0.98-1.62
condition of the apartment	1225	57.1	2.07	1.74-2.47	1.19	0.95-1.50
"Apartment-related social components"[†]						
relationship with neighbours	951	44.4	2.30	1.93-2.75	1.46	1.19-1.80
perceived noise from neighbours	993	46.3	2.26	1.90-2.70	1.32	1.07-1.64
rent / mortgage	936	43.7	1.49	1.25-1.77	1.16	0.95-1.42
"Suitability for children"[†]						
institutionalised day care	101	4.7	2.22	1.41-3.49	1.45	0.84-2.48
"Cultural and social life"[†]						
clubs / associations in neighbourhood	455	21.2	1.93	1.55-2.41	1.28	0.99-1.65
"Community services"[†]						
cleanliness of the surroundings	1050	51.0	2.16	1.82-2.58	1.24	0.99-1.56
"Infrastructure"[†]						
medical supply	412	19.2	1.49	1.19-1.86	1.23	0.93-1.62
facilities for daily shopping	615	28.7	1.33	1.10-1.61	1.22	0.96-1.54
"Leisure time"[†]						
supply/security of sidewalks	640	29.9	2.01	1.65-2.44	1.21	0.96-1.54

*   *Compared with "not improved".* [†] *Grouping and labels derived from the factor analysis as shown in*
figure 3-1. [‡] *Adjusted for all indicators presented and for sex, age, household composition and*
income, education and type of moving

Environmental housing quality indicators and improved self rated health in subgroups

Subsequently, we investigated if the associations with the two "environmental" indicators "perceived air quality" and "location of the building" were modified by the main reason to move, the type of moving or whether participants had moved from a multiple dwelling into a single family home or had become a house owner. In table 3-3, the results of the stratified logistic regression analyses are presented.

Table 3-3:   *Associations between an improvement in self rated health since having moved and an improved satisfaction with the "perceived air quality" and the "location of the building" in different subgroups of movers*

	nr.	Perceived air quality			Location of the building		
		OR*	95% CI	$\chi^{2\,\dagger}$	OR*	95% CI	$\chi^{2\,\dagger}$
Total sample	1961	1.58	1.24-2.01		1.58	1.28-1.96	
Moved into single family home	1944						
yes	268	3.28	1.46-7.38		1.16	0.61-2.22	
no	1676	1.44	1.11-1.87	p=0.06	1.69	1.34-2.12	p=0.28
Type of moving	1961						
out of the city	1028	2.27	1.61-3.20		1.56	1.15-2.12	
within the city	933	1.19	0.83-1.70	p=0.01	1.58	1.16-2.15	p=0.95
Main reason to move	1825						
environmental reason‡	390	2.28	1.20-4.31		1.89	1.05-3.39	
other reasons	1435	1.38	1.05-1.83	p=0.16	1.58	1.23-2.02	p=0.58
Type of moving and main reason to move	1825						
Moved out of the city							
environmental reason‡	230	4.58	1.76-11.89		1.89	0.84-4.26	
other reason	739	1.81	1.22-2.69	p=0.08	1.55	1.09-2.22	p=0.66
Moved within the city							
environmental reason‡	160	1.08	0.37-3.13		3.63	1.20-11.03	
other reason	696	1.18	0.77-1.80	p=0.88	1.48	1.03-2.12	p=0.13

* *Final logistic regression model adjusted for sex, age, household composition and income, education, type of moving and all indicators presented in table 2.* † χ^2 *test for heterogeneity of estimates.* ‡ *Including the categories "environment" (for example, noise, traffic, not enough green) and "neighbourhood" (for example, suitability for children, problems with neighbours, dirt)*

Among the less than 15% of participants who had moved into a single family home, improved self rated health was more strongly associated with a more favourable assessment of the "perceived air quality" than among the remaining subjects. Becoming a house owner did not alter the associations materially, moreover the respective subgroup was small (results not shown).

"Type of moving" also modified the association with "perceived air quality": Among subjects having moved out of the city, the odds ratio between an improvement in self rated health and a higher satisfaction with this indicator was twice as high as among within-city-movers. This association was also stronger in participants who had moved mainly for environmental reasons but it was still statistically significant among those who had moved for other reasons.

When the analyses were stratified by type of moving and by main reason to move, the association between improved self rated health and a higher satisfaction with the "location of the building" was found in both types of movers, being slightly stronger in those who had moved within the city for environmental reasons. The association with an improved assessment of air quality on the other hand was only found in subjects who had moved out of the city. It was stronger in those having moved out of the city for environmental reasons but still remained significant in those with other reasons.

3.4 Discussion

Our results show that even in an economically well-to-do country like Switzerland, a higher satisfaction with the environmental quality of the new housing surroundings was associated with an improved wellbeing of movers when other dimensions of housing quality and potential sociodemographic confounders were taken into account even if the subjects hadn't moved for environmental reasons.

3.4.1 Importance of different dimensions of housing quality

We found that the satisfaction with the environmental housing quality, with the apartment and with the apartment-related social environment were more strongly associated with wellbeing, than were infrastructure indicators, the suitability for children, and the cultural and social life. Only a limited number of prior studies are available to compare these findings to. Most of them either focused on specific aspects of housing quality and health[41, 42] or used different outcome or exposure measures. Van Poll also found that subjective health (based on reported symptoms) was associated with dwelling satisfaction but not with neighbourhood satisfaction.[63] Lawton found rather similar associations between a perceived positive change in one's life (including health) and interviewer-rated ambience of the dwelling, dwelling maintenance and neighbourhood quality.[63]

The relative importance of different dimensions of housing quality varies probably across different cultures and social groups. Even though our finding seems plausible, the issue remains complex. Some of the dimensions and respective indicators of housing quality in our study whose associations with an improved self rated health were borderline significant would certainly deserve further investigation. It is also interesting to note that an improved relation with neighbours and less perceived noise from neighbours, reflecting the apartment-related social environment, seem to be just as important for an improved wellbeing of movers as the physical characteristics of the apartment itself.

3.4.2   Environmental housing quality indicators

Since our research question focused on the environmental housing quality we explored this dimension in more detail. The perception of the two "environmental" indicators "location of the building" and "perceived air quality" probably differed somewhat between individuals and we suppose they stand for two slightly different aspects of the residential environment. Nonetheless, both indicators were clearly grouped in the same factor "environment".

The rather global environmental indicator "location of the building" seems to reflect different aspects in the more immediate neighbourhood since next to the association with environmental indicators it was also weakly correlated with e.g. suitability for children, commuting related indicators, supply with infrastructure, and social characteristics of the neighbourhood (however, correlation coefficients were all below 0.3). This indicator was associated with an improved wellbeing of movers irrespective of the destination or reason of moving. The immediate neighbourhood therefore seems to be of general importance even though we found an indication that a positively perceived change in this indicator may be more important for an improved wellbeing among subjects who had moved within the city for environmental reasons.

The indicator "perceived air quality" does not entirely reflect the objectively measurable air quality. This indicator should rather be understood as a qualitative evaluation which was also associated with other indicators of environmental quality such as greenness of surroundings as well as noise and negative effects of traffic. It can therefore also be interpreted as a proxy for "environmental quality" in a more general sense. The restriction of the association between the "perceived air quality" and wellbeing to subjects who had moved out of the city is therefore of relevance for the ongoing debate on the reasons of suburbanisation in Switzerland.[57] Since no information on objective measures of environmental quality was available, we cannot determine from our data whether this reflects a real difference in the environmental quality between city and surrounding areas or just different perception between within- and out-of-city-movers. That the former is true is suggested by the fact that air pollution was rather uniform within the city of Basel[64] while somewhat lower concentrations were found at sites surrounding the city.[65] Thus, the observation of a stronger association among subjects who had moved out of the city supports our interpretation of a change in satisfaction with the "perceived air quality" as reflecting a real difference in the environmental quality, since the achievable degree of perceived improvement was likely to be bigger among those subjects.

That among the environmental indicators, an improved satisfaction with "perceived air quality" was most predictive of an improved self rated health certainly also reflects the current political and public debate. In Switzerland, air quality has been a main issue for several years while e.g. noise has received less public attention so far.

3.4.3 Type of moving

Having moved into a single family home also increased the association between an improved self rated health and a better assessment of "perceived air quality". However, the increase was only borderline significant and we suspect that having moved into a single family home was less influential than having moved out of the city, since only 14.9% of the total sample and 23.1% of the out of city movers actually moved into a single family home.

3.4.4 Methodological consideration

A number of aspects are of relevance for the interpretation of the results. First, it should be noted that even though the response rate was not particularly high participation bias does not seem to be a problem in our study. In most cases, non-participation was due to technical reasons. A comparison with data from the statistical office of the canton of Basel regarding sex, age, nationality and type of moving showed that our sample was representative of the base population except for non-Swiss participants who were slightly underrepresented, especially in the older age groups. The educational level seemed to be rather high compared to the general Swiss population.[66] However, it was to be expected that the sample might contain more subjects with a higher socioeconomic status. Inhabitants with lower education and income, especially foreigners, are more likely to have difficulties in finding new residences.[67-69]

As shown in the factor analysis, certain clusters of interdependent housing quality indicators were found in our data. Collinearity can lead to difficulties in separating the effects of individual indicators in a multiple regression analysis. This is a possible explanation for the observed weakening of the associations in the multivariate analyses. However, only one correlation between the 14 indicators in the final regression models exceeded 0.50 (suitability and comfort of the apartment: 0.51) and only 6 were above 0.30. Nevertheless, 5 associations remained significant in the multivariate analyses.

The simultaneous collection of the information on former and present housing quality may be a source of measurement error leading to an overestimation of the associations presented if recall bias has the same direction for prior housing quality and prior self rated health. Marans concluded however that biases introduced by dissonance reduction (i.e. the tendency to avoid conflict between past action and current feelings) were not very large and that they applied rather to general evaluations than to assessments of specific attributes as presented here.[70] Francescato proposed to use relative degrees of satisfaction as done in this study.[71] Moreover, since the recall period in our study was relatively short, we consider it as unlikely that recall bias may have been a major source of error.

Whereas "self rated health" is a useful summary measure to study a more general concept of health, its global and subjective character does not allow to determine which aspects of wellbeing - physical, psychical or social - are most affected by perceived improvements in the housing quality. Unfortunately, additional information on objective health measures to explore this issue further were not available. Thus, we also could not control for a change in morbidity or for a decline in functional ability in our subjects, factors which have been shown to influence self rated health.[53, 72, 73] However, we consider changes in objective health rather as a possible intermediate step than as a potential confounder having influenced the choice of the new residence, particularly since only a small proportion of participants reported a deterioration in self rated health and less than 1 per cent mentioned health and/or ageing as main reason to move.

Even though the cross-sectional design of this study does not allow for causal inference, it must be noted that we assessed the change in self rated health and the change in perceived housing quality since having moved, reflecting thus a time interval. So far, only very few longitudinal studies on changes in the satisfaction with housing quality in a broader sense and subsequent changes in wellbeing have been carried out. The few available studies included different age groups and used different dependent variables and indicators for housing quality than our study.[11, 46-48] Despite our own and a few other results suggesting a causal relationship of housing quality on wellbeing, additional longitudinal studies with population based samples covering wider age ranges and using more detailed sets of indicators for housing quality are needed to further elucidate temporality.

Mackenbach and co-workers showed in a cross-sectional study that housing problems decreased the probability of excellent self rated health.[44] They suspected that this association might be an artefact of a propensity to complain because they used few general indicators to measure such a complex construct as "housing quality".[74] It seems unlikely that the specific and plausible patterns of the reported associations in our study are merely an artefact of general negativism since we used a large set of indicators and individual answering patterns varied substantially. It must also be kept in mind that all subjects in our study had moved and had done so within the same time frame. Therefore, the results cannot be confounded by a "honeymoon" reaction following a change in residence within a subsample. On the other hand, a general improvement in life satisfaction following voluntary moving which might be present in the whole sample cannot explain the heterogeneity of associations across various subgroups of movers. Certainly, subjective assessments of the environment are influenced by personal characteristics as well as by beliefs, emotions, and behavioural intentions.[71, 75] The individual response to an adverse environmental situation depends also on appraisal of the source and on e.g. controllability and predictability of the stressor.[76] Nevertheless, if the impact of housing quality on residents' wellbeing is the target of interest, individual perception is the driving force and should therefore be of interest despite these limitations, unless one is willing to state that "the expert knows better".

3.5 Conclusions

Our results add to the understanding of a complex issue even though we could not entirely clarify which factor of the housing environment was most influential for an improved wellbeing of movers. However, we showed that perceived environmental quality is an important predictor of wellbeing of citizens. Moreover, the significant associations between perceived improvements in the two environmental indicators "location of the building" and "perceived air quality" and an improved wellbeing in participants who had not moved for environmental reasons certainly deserve attention. Further longitudinal studies on changes of wellbeing should therefore take moving, motivations to do so and subsequent changes in satisfaction with environmental housing quality into account.

Annex:
Question verbatim of the 14 housing quality indicators in the final model (table 3-2)

We are now going to name different aspects regarding the housing quality and quality of life and ask you again to give grades between 1 and 6, first for your present and afterwards for your former residence. 1 is the worst, 6 the best grade, 4 is sufficient, half grades may be given.

- air quality (present) / (former)
- noise from neighbours (present) / (former)
- cleanliness of the surroundings (present) / (former)
- comfort of the apartment (size, facilities) (present) / (former)
- level of the rent or mortgage (present) / (former)
- suitability of the apartment referring to your needs (present) / (former)
- condition of the apartment (present) / (former)
- location of the building referring to your needs (central or quite, green surroundings etc.) (present) / (former)
- relationship with neighbours (present) / (former)
- facilities for daily shopping needs close by (present) / (former)
- medical supply, hospitals, pharmacies (present) / (former)
- institutionalised day-care (present) / (former)
- clubs/associations in the surroundings regarding your needs (present) / (former)
- supply and security of sidewalks (present) / (former)

4 The first years of implementation of the Swiss National Environment and Health Action Plan (NEHAP): Lessons for environmental health promotion[*]

Abstract

The National Environment and Health Action Plans (NEHAPs) are a novel attempt to integrate environmental protection and health promotion in political programs. Throughout Europe, about 40 NEHAPs have been developed so far. The Swiss NEHAP was among the first to be developed in an industrialized country. We discuss the Swiss NEHAP and draw first conclusions on the development and implementation process of such programs, using illustrative examples of other European NEHAPs. The strengths of the Swiss NEHAP lie in the formulation of specific targets in selected areas, its approach as a environmental health promotion program, and its comprehensive evaluation. Weaknesses in most NEHAPs are the lack of involvement of the general public and of the economic sector and the absence of an implementation strategy along with adequate financing.

[*] *Published as: Kahlmeier S, Künzli N, Braun-Fahrländer C: The first years of implementation of the Swiss National Environment and Health Action Plan (NEHAP): Lessons for environmental health promotion. Soz Praventivmed 2002: 47:67-79 (including 3 commentaries, see chapters 4.4, 4.5 and 4.6).*

4.1   Environmental health promotion

Almost 150 years ago the link between environment and health was formally recognized after a cholera outbreak in London.[27] In the course of time, environmental health developed from a synonym for "sanitation" at the beginning of the century to a public health issue. The environmental movement in the middle of the twentieth century supported this development with its concern for environmental pollution.[77] The recognition of the importance of the subject for public health which followed later on was also enhanced by major environmental health disasters.[30] In Switzerland, the Schweizerhalle-accident had a major impact on public attitude towards environmental pollution and health.[78] The promotion of environmental health in a more integrated way developed by the end of the twentieth century. Based on the WHO-report "Our Planet, our Health"[5] prepared for the Earth Summit in Rio de Janeiro in 1992, a variety of environmental health promotion measures was outlined in Agenda 21.[33] The subject was further developed and substantiated as part of the practical implementation of sustainable development at the European WHO Conference on Environment and Health in 1994.[7]

The novelty of these concepts was the explicit linking of the formerly separated areas of environmental protection and health promotion[79,80] and a broadened concept of "health" defined as a dynamic process.[14] This concept encloses both individual behaviour and conditions stating that political, economic, social, cultural as well as environmental factors all are influential for health and wellbeing. Therefore, the prerequisites of health cannot be ensured by the health sector alone but health must be integrated into the planning and implementation processes of the different administrative sectors and levels in order to create a supportive environment.

Based on the European Action Plan,[7] about 40 National Environment and Health Action Plans (NEHAPs) have been developed which seek for the application of these concepts. While the program has an important impact in eastern European countries[81,82] positive experiences from the western European region are more rare. The Swiss NEHAP[40] was among the first to be completed in an industrialized country. As external evaluators of the Swiss NEHAP, we will highlight and discuss its strengths and weaknesses and draw first conclusions on the development and implementation process of such programs, illustrated by selected examples of other European NEHAPs.

4.2   The Swiss National Environment and Health Action Plan

The Swiss NEHAP was developed from 1995 to 1997 as part of the Swiss Action Plan for Sustainable Development.[40] The Federal Office of Public Health (FOPH) and the Swiss Agency for the Environment, Forests and Landscape (SAEFL) jointly guided the development process (see figure 4-1). A concept working group was formed consisting of representatives of the cantons and municipalities and campaigning NGOs as well as representatives from the science sector and of professional groups. This concept working group formulated the central idea of the Swiss NEHAP: the promotion of health and wellbeing of all people in a healthy environment.

4.2.1 Problem analyses and priority setting

Even though in Switzerland basic environmental requirements for good health such as the supply with safe water and food, waste disposal or occupational safety are mostly ensured, there are still areas which need improvement.[40,83] Therefore, at first a problem analysis was carried out to identify priorities. Legislation and existing programs where taken into account to avoid duplication: Areas like sanitation or chemical safety, in which the existing measures were considered to be sufficient, were not included. Subsequently, 17 topics were rated by each member of the concept working group according to the following criteria: impact on ecology and health, scientific evidence of the relevance of the problem and of a causal association, long term negative effects, economic burden, political sensibility, perception in the society, and relation to the European program. Another leading question in this process was on which topics the link between environment and health could be communicated easily.

Figure 4-1:   *Development Process of the Swiss National Environment and Health*
              Action Plan (NEHAP) and participating institutions (Fed. Off. = Federal
              Office, NGO= nongovernmental organisation)

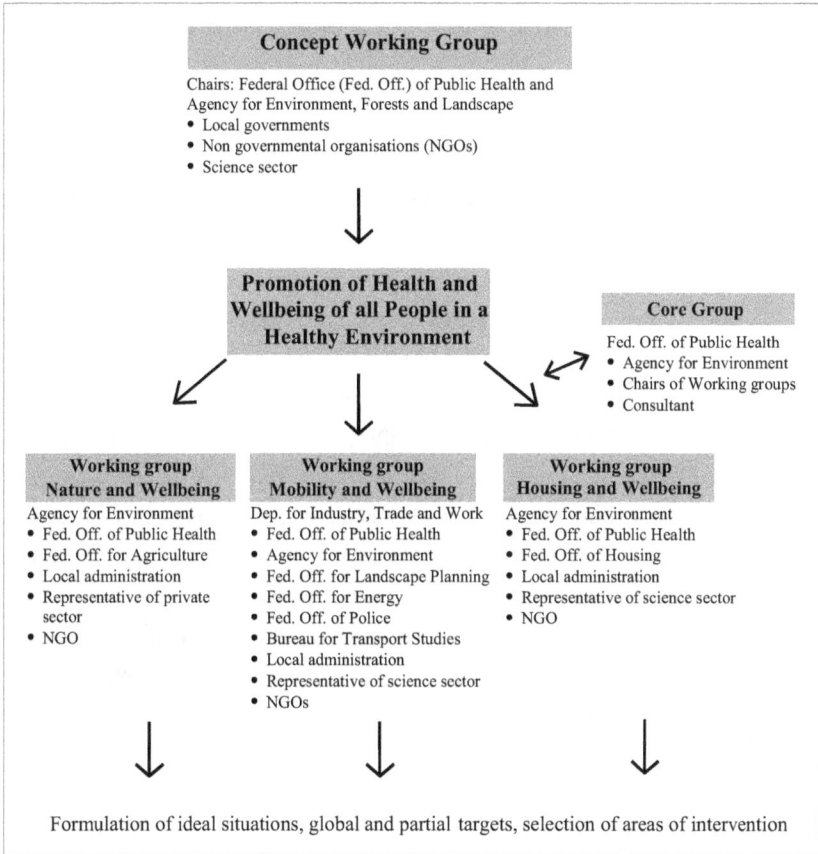

The ranking of the concept working group members resulted in the choice of the following three areas:

- Nature (i.e. agriculture and nutrition) and Wellbeing,

- Mobility and Wellbeing,

- Housing and Wellbeing.

These three areas are not separate fields. In figure 4-2, the complexity of the interactions between them is illustrated (modified from[40]).

Finally, an interdisciplinary working group was formed for each of the areas of the NEHAP which had to formulate specific targets and measures (see figure 4-1). Subsequently, a draft of the NEHAP was discussed in hearings with various interest groups.

4.2.2 Targets and measures

An ideal situation was laid down for each area as a starting point for the formulation of a global target which was further specified in partial targets and areas of intervention (see table 4-1)[40]. The targets and measures were formulated wherever possible in such a way that they will have an impact both on health and environment. E.g. the promotion of human powered mobility, one of the partial targets in the area "Mobility and Wellbeing" presented in table 4-1, is on the one hand a means to reduce detrimental environmental effects of motorized traffic like emissions or space consumption. On the other hand, a doubling of ways made by bicycle would lead to more people exercising on a regular basis. Thus, the promotion of human powered mobility is an ideal measure on the way to the vision of the NEHAP in this area: A mobility enhancing human wellbeing while conserving the environment. To achieve this partial target, it is not only planned to rise public awareness but to improve at the same time the conditions for cycling through e.g. landscape planning or incentives by employers ("Areas of intervention", table 4-1).

Figure 4-2:   Interactions between the three areas of the Swiss National Environment and Health Action Plan (modified from[40])

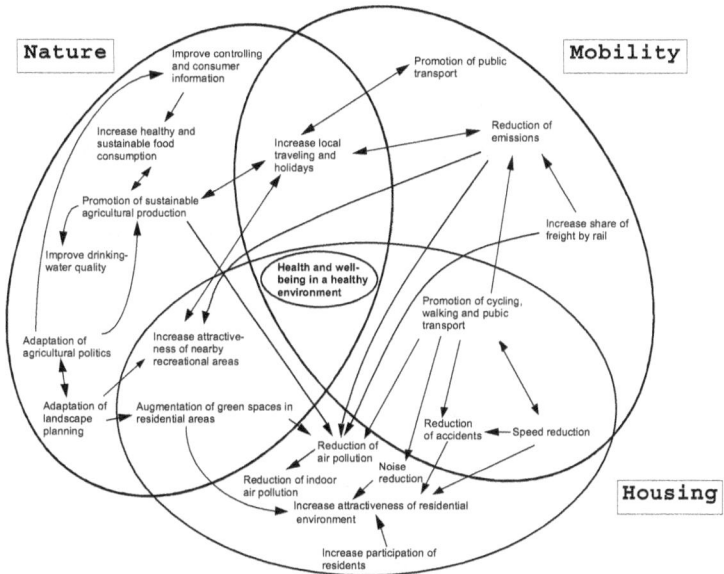

4.2.3   Implementation

The program is translated into action since 1998 under the guidance of the FOPH. The Swiss NEHAP is aimed at being effective in itself but at the same time, it is embedded in the context of other policies and programs which have already been initiated. It was intended to complement existing activities with regard to environmental health promotion and to serve thereby as an instrument to intensify intersectorial cooperation. As first step of the implementation, working groups consisting of the concerned Federal Offices and of the local authorities were established to coordinate the activities and to build a structural network at the national and local level. In November 2001, the NEHAP-project database contained information on 48 projects. 35 % of these projects were started because of the NEHAP, in the remaining the FOPH is involved in the project management or financing.

4.2.4 Evaluation

Evaluation should be an inherent part of every health promotion program.[17] The evaluation concept for the Swiss NEHAP developed in 1997 is based on a goal oriented, user focused approach.[35] The planning and implementation process as well as outcomes and impacts are studied. The continuous evaluation of the implementation is based on a series of interviews, document analysis, and the aforementioned NEHAP-project database. Impact models were formulated as basis for the choice of indicators to assess the effectiveness of the implementation in relation to the targets. A baseline assessment of these indicators was carried out in 1999 against which progress can be measured later on (http://www.unibas.ch/ispmbs/dienst/e/edie301.htm).

4.3   Strengths, weaknesses and first conclusions

The Swiss NEHAP is innovative in a number of aspects: First of all, the aim was to create a promotion program with its own specific targets at the interface of environment and health. This is a first distinction to other European NEHAPs such as the Austrian, which mainly represents an overview of existing legislation, measures and programs.[84] Another difference to most NEHAPs is the positive, health-based approach focusing on "wellbeing" instead of indicators of illness. Further, the majority of the Swiss targets were quantified, stating which level of improvement shall be achieved until when (table 4-1). Exceptions were made in areas which are politically sensitive (like the time frame concerning the impact threshold levels for air pollution) or which still lack scientific basis (e.g. definition of "attractiveness of housing environment"). Obviously, this quantification facilitated the development of an evaluation concept considerably.[85] Accordingly, in most NEHAPS the assessment of the implementation and goal attainment is only mentioned in a very general way or not at all. So far, only in very few countries apart from Switzerland, an evaluation has been put into practice, e.g. in Hungary.[86]

On the other hand, due to restricted resources for the development of the Swiss NEHAP, the analysis of existing programs, legislation and administrative structures was quite limited. Additionally, the participation in the working groups was solely based on voluntariness and decisions were not always transparent. Another weakness is the lack of involvement of the economy and the general public. While for example in Poland, stakeholders of various economic sectors were involved in the priority setting process[87] or in the Ukraine, a separate chapter in the NEHAP was dedicated to public participation,[88] Switzerland as most other countries did not provide specific measures to involve these groups. This contradicts one of the basic principles of health promotion programs, i.e. the participation of the ones affected[14] and leads to non-collaboration of a key partner: the economy.[89]

However, the lack of a comprehensive implementation strategy as part of the action plan is probably the most important weakness of a number of NEHAPs. In most NEHAPS, e.g. the need to intensify the collaboration between various departments and administrative levels to achieve improvements in the environment and health area is emphasised. Yet, only a few NEHAPs state how this intention shall be put into practice, like e.g. the Bulgarian: An Interagency Steering Committee, jointly guided by the Minister of Health and the Minister of Environment, is responsible for the coordination and continuous control of the implementation in all concerned departments.[90] Another positive example is Poland which worked out a separate implementation program.[87] The lack of such an implementation strategy involves the risk of inefficiency, actionism, arbitrariness in the choice of partners, vague communication, and thus ineffectiveness. Additionally, it impedes a systematic evaluation of the implementation process. Also in Switzerland, the implementation had not been addressed adequately in the action plan itself. However, as a consequence of the process evaluation revealing this fact, an implementation strategy has been developed recently.[91]

The separation of the NEHAP- and the Agenda 21-process at the Rio-Conference, which continued on the national level, turned out to be another powerful hindrance.[92] Despite international efforts to integrate the association between environment and health into decision making and policy formulation,[93] in daily business the two areas still operate mainly within divided structures in most European countries. Therefore, the formulation process of the NEHAP served as cornerstone for the discussion and transfer of knowledge between hitherto mostly separated disciplines and thus as a starting point to pull the pieces together. However, in Switzerland the FOPH alone was assigned with the implementation. Since the FOPH does not have the authority to issue directives to the other involved administrative bodies, it depends on their non-material as well as material support. Even though the process evaluation showed that the working groups served their purpose well in ensuring the involvement of the relevant partners, it became also apparent that the identification with the project and the respective role in it as well as the degree of cooperation still depended strongly on the individual representatives. Hence, for the establishment of a stable environment-and-health-network independent of involved individuals, further effort, resources, and time are needed as well as a comprehensive implementation strategy tackling the inherent centrifugal forces steming from the complexity of the field (see figure 4-2).

Thus the greatest challenge in the implementation of this in principal valuable framework will be to ensure the link between health and environment on a structural level beyond an intersectorial development phase to build a real and long-term stable alliance.[15,16] An implementation strategy translating the action plans into an "action process" and adequate financing are crucial, as well as the involvement of the public and the economy. Finally, systematic evaluations would add to the effectiveness and credibility of the NEHAPs.

Table 4-1: *Ideal situations, targets and areas of intervention of the Swiss National Environment and Health Action Plan*[40]

Agriculture, nutrition and wellbeing	Mobility and wellbeing	Housing and wellbeing
Ideal Situation		
Nature and landscape are conserved and used in such a way that there is a harmonious balance between human wellbeing and the conservation of natural resources.	Mobility is applied in such a way that it enhances our wellbeing while our environment is conserved.	The quality of settlements is improved in such a way that it promotes our wellbeing and allows individual active involvement.
Goal		
By 2007, ¾ of the Swiss population will be in a position to consume healthy, balanced and enjoyable food, thus contributing to sustainable agriculture.	By 2007, current adverse impacts of motorised mobility will be reduced by a significant reduction in adverse emissions, and by increase in proportion of non-motorised mobility.	By the year 2007, healthy and environmentally adequate housing will be assured in 90% of all residential areas.
Targets		
• By 2002, 80% of the population will know how to eat healthily and in harmony with seasons and that their consumption patterns influence agricultural production and landscape. • By 2007, nearly 100% of the agricultural soils will be used according to the principles of integrated pest management or organic production (OP), proportion of OP > 30%. • By 2007, 70% of the available meat will be from species appropriate and livestock-friendly production. • By 2007, nitrate content of 99% of all drinking water collectors will be <40 mg/l. • By 2007, 90% of all agricultural and related business will have standardised quality control systems; positive declaration/reproducible production pathways are the rule.	• By 2002, 80% of the population will know about the interactions of motorised traffic, emissions and adverse impacts on human health. • Emissions of motorised traffic will be reduced to such an extent that the impact threshold levels of the Ordinance on Air Pollution Control can be respected. • By 2007, the proportion of journeys by bicycle will have doubled for commuting, shopping and leisure as compared to 1995 (7%, 5% and 7%, respectively).	• By 2002, 80% of the population will be well informed about indoor air pollution and able to take adequate measures. • By 2002, a speed limit of 30 km/h will be introduced in 70% of urban and peri-urban residential areas. • By 2000, no-one will be exposed to involuntary passive smoking in the workplace, means of public transport and public buildings. • By 2007, residential areas will have structures to encourage active involvement in neighbourhood life. Planning interventions will create conditions allowing adequate presence of small manufacturers, jobs (esp. supply), leisure and services.
Areas of intervention		
1. Information/education/training of all partners of the population concerning environmentally adequate and healthy food (e.g. campaigns, schools) 2. Intensification of contacts between consumers and producers/farmers (e.g. direct marketing) 3. Implementation of the Swiss Agrarian Reform 4. Establishment of labelling and quality control systems for agricultural products and the production of such, in order to enhance truth-in-packaging for consumers	5. Promotion of public awareness of mobility related issues of safety and health (e.g. schools, campaigns) 6. Reassignment of roads and improvement of traffic flow to promote non-motorised traffic 7. Incentives to transfer traffic to public transport and bicycle (e.g. parking, access to public transport) 8. Protection of the alpine region by reducing motorised traffic (e.g. Alp Initiative, tourism) 9. Reduction of emissions from motorised traffic	10. Promotion of public awareness of indoor air pollution and adequate behaviour (e.g. schools, campaigns) 11. Promotion of 30km/h speed limit (e.g. streaming of legal procedure, information) 12. Prevention of nuisances by passive smoking 13. Enhancing attractiveness of housing environment (e.g. meeting places) 14. Upgrading of nearby recreational and green areas within urban residential areas

4.4   Commentary I: Towards assessing effects of National Environmental Health Action Plans

by Dr Michal Krzyzanowski, *WHO Centre for Environment and Health Bonn Office*
DISCLAIMER: The views expressed in this commentary are those of the author and do not necessarily represent the decisions or stated policy of the WHO.

Following the 2nd Ministerial Conference on Environment and Health in Helsinki, in 1994, most of the Member States of the WHO European Region prepared NEHAPs. Ministries of health, public health agencies and professionals were the driving force in this work. However, an important feature of all the programmes was active involvement of environmental agencies, and of the other sectors, contributing to the quality of the environment and its potential health impacts. In many countries, the NEHAP preparation provided the first opportunity for the direct collaboration and exchange of information between these sectors. The appreciation of the importance of the strong intersectorial collaboration lead to the selection of the "Action in partnership" as the leading theme of the 3 rd Ministerial Conference held in London in 1999.

A comprehensive assessment of the extent of implementation of the NEHAPs on the international scale has not been conducted yet. The paper on the Swiss NEHAP is one of few examples of a national evaluation of the NEHAP implementation. The authors point out the usefulness of the definition of the quantified targets in the NEHAP design, allowing assessment of progress in the programme implementation. The focus on "well-being" is an important feature of the programme, underlying the need to work on the environmental improvements not only when the poor environmental quality increases the risk of clinically recognisable illness. While the programmes aimed at strong intersectorial collaboration, the authors assess the implementation of this objective as limited. Somewhat discouraging is the observation that only 10 % of projects included in the NEHAP data base have been initiated because of the NEHAP.

The occasion for the evaluation of NEHAP implementation in some other countries provide the Environmental Performance Reviews, completed by UNECE, with WHO contribution (United Nations Economic Commission for Europe s.d.). Observations from several countries of Central and Eastern Europe, and Central Asia confirm the conclusion of the Swiss paper, stating that the lack of implementation strategy is an important weakness of the NEHAPs in those countries. The general objectives to reduce the risks from hazardous exposures are not translated into operational programmes. Lack of operational targets and instruments to measure extent of their achievement also reduces the ability to evaluate the NEHAPs implementation. There is a risk that the NEHAP documents, even those approved by the legislative bodies in the countries, will remain on paper and will not contribute to the improvement of health of the people.

Public health professionals should recognise the value of the work performed to compile the NEHAP documents and use it as a basis of their actions to promote healthy environment. Finding of measurable improvements in environmental quality and reduction of health risks that may result from the projects implemented in the framework of NEHAPs is the best argument for further actions and for support the NEHAPs. While many of the actions must be implemented outside of the public health sector, the assessment should be the responsibility of public health agencies and professionals. It requires assessment of changes in the exposures affecting health, as well as in the health aspects associated with environment quality. Several of such measures are readily available. However many of the health or environment characteristics being addressed by the NEHAPs are not measured or measured with poorly standardised or validated methodologies. Development of the Environmental Health Indicator system, coordinated by WHO, aims at harmonisation of efforts to develop the necessary assessment tools and to adjust them to the needs of policy setting and its evaluation (World Health Organization 2000). Contribution of public health professionals to the development and implementation of the system may be one of their important tasks to the increase of NEHAPs effectiveness and visibility.

References

World Health Organization (2000). Environ-mental health indicators: development of a methodology for the WHO European Region. Interim Report November 2000. Copenhagen: WHO, Regional Office for Europe. (EUR/00/5026344). http://www.who.nl/download/doc45/ehi-report.pdf (acc. Dec. 2001)

United Nations Economic Commission for Europe (s.d.). Environmental performance reviews. S.l.: S.d. http:/www.unece.org/env/epr/(acc. December, 2001).

4.5   Commentary II: Environment and health: from national policies to global initiatives

By Francesco Forastiere MD PhD, head of the Analytical Epidemiology Unit at the Department of Epidemiology, Lazio Regional Health Authority, Rome

In this issue of the journal, Kahlmeier and collaborators provide an interesting update of the efforts to implement the National Environment and Health Action Plan (NEHAP) in Switzerland, within the Swiss Action for Sustainable Development. The basic idea of the NEHAP is that health is the outcome of all the factors and activities acting upon the lives of individuals and communities. Various sectors of the society, not only the health sector, have to be involved in planning, financing, and taking care of the different issues with a potential impact on health. This concept has been reiterated during the Third Ministerial Conference on Environment and Health (World Health Organization 1999). A good application is the Charter on Transport, Environment and Health as a framework for measures to facilitate the integration of health issues in decisions, planning and investments affecting transport and mobility. Environmental monitoring, quality assurance, epidemiologic expertise, health impact assessment, work in partnership, professional experience in risk communication, all are the key elements for a success.

The practice is always more difficult than the theory, however. It has been already indicated that the field of environmental health on one side and that of public health on the other side have repeatedly found themselves isolated and separated (Kotchian 1997). Environmental agencies often neglect their public health responsibilities and public health agencies abdicate their environmental responsibilities under the pressure of the "health market". Even the simple role of advocacy of the public health agencies to demand structural changes in order to implement primary prevention measures is often forgotten. To "ensure a link between health and environment on a structural level", to "translate the action plans in an "action process with adequate financing", as the authors stress, are urgent needs not only for Switzerland but also for many European countries. This is the difficult world of national policies. "Globalisation", however, is the new word that defines the current era. It has several implications for those involved in environment and health issues. Some of these implications, in particular with regards to epidemiology, have been recently reviewed (Hertz-Picciotto & Brunekreef 2001). Within this context, there are two menaces for the health status of our world: wars and ecological changes. The first is immediate while the second requires some time to fully express all its impact. The question is: should we bother of them? During these days of war when the fear of terrorist attacks undermines our lives, "collateral damages" kill innocent people, UN pleads for break in bombing in Afghanistan (Ahmad 2001), all the potential health effects directly and indirectly associated with war are difficult to be foreseen. Many of the indirect effects will take place through environ-mental destruction, use of biological and chemical weapons, limit in the use of natural resources, mass mobilisation, all leading to drought, famine and humanitarian disaster (Horton 2001).

The adverse health consequences of climatic change have been made clear by an international scientific body (Inter-governmental Panel on Climate Change 2001): the warming has already begun, changes in physical and biological systems are apparent across all continents, a temperature rise in this century has been foreseen (McMichael 2001a; McMichael 2001b). Large-scale environmental changes are now under way. All these changes have great consequences for the sustainability of ecological systems, for food production, economic activities, and human health. As McMichael (2001a) has clearly stressed, "… in the long run, it is the conditions of social and natural environments that set the limits to human health and survival and that determine the patterns of disease". Only radical changes in energy systems, and in setting economic and social priorities could reverse this process. Unfortunately, the president of the most developed nation, with the greatest responsibility for the green-house effect, refused even small changes under the Kyoto Protocol (McMichael 2001a). In the mean time, we all know that air pollution from current fossil fuel use for transportation, industry and housing is killing millions throughout the world (Künzli et al. 2000; Cifuentes et al. 2001).

In conclusion, difficulties at national level to implement integrated policies for environment and health will certainly require effort, coordination, and public participation. During these days, however, we cannot ignore that the large-scale impacts induced by wars and ecologic shifts need to be addressed by those interested in public health.

References

Ahmad K (2001). UN pleads for break in bombing in Afghanistan. Lancet 358 : 1352.

Cifuentes L,Borja-Aburto VH,Gouveia N, Thurston G,Davis DL (2001). Climate change: hidden health benefits of greenhouse gas mitigation. Science 293 : 1257–9.

Hertz-Picciotto I,Brunekreef B (2001). Environmental epidemiology: where we've been and where we're going. Epidemiology 12 : 479–81.

Horton R (2001). Public health: a neglected counterterrorist measure. Lancet 358 : 1112–3.

Intergovernmental Panel on Climate Change (2001). Climate change 2000: third assess-ment report, vol. 2: Impacts, vulnerability and adaptation. Cambridge: Cambridge University Press.

Kotchian S (1997). Perspectives on the place of environmental health and protection in public health and public health agencies. Annu Rev Public Health 18 : 245–59.

Künzli N,Kaiser R,Medina S , et al. (2000). Public-health impact of outdoor and traffic-related air pollution: a European assessment. Lancet 356 : 795–801.

McMichael AJ (2001a). Climate change and health: information to counter the White House Effect. Int J Epidemiol 30 : 655–7.

McMichael AJ (2001b). Global environmental change as "risk factor": can epidemiology cope? Am J Public Health 91 : 1172–4.

World Health Organisation (1999). Third Ministerial Conference on Environmental and Health. London, 16–18 June 1999. http://www.who.it. (acc. Dec. 2001).

4.6 Commentary III: A joint effort in the field of environment and health

By Dirk Ruwaard, Public Health Division, National Institute of Public Health and the Environment, Bilthoven, and

Pieter G.N. Kramers, Department for Public Health Forecasting, National Institute of Public Health and the Environment, Bilthoven

One important observation is that in the past measures related to environmental protection contributed a lot to enhance health at the individual and population level. Can we expect additional health benefits in industrialised countries nowadays? The Swiss National Environment and Health Action Plan (NEHAP) was among the first to be developed in an industrialised country. We fully support the statement that health cannot be ensured by the health sector alone. Health must be integrated into the planning and implementation processes of the different administrative sectors and levels in order to create a supportive environment. To develop such joint efforts, for instance in the field of health and environment, it is a prerequisite to create a situation of mutual benefit. The targets and measures must have an impact both on health and environment, which seems to be part of the Swiss NEHAP. So far so good.

However, the paper of Sonja Kahlmeier et al. also raises questions. We will highlight three topics. The first one relates to the selected areas and their underlying concepts. From 17 areas, the working group members selected the following three: nature and well-being, mobility and well-being, and housing and well-being. How was this selection made? The paper gives seven criteria on which the choice was based, but has not made clear how these criteria of very different sorts were weighed in order to make the final selection. In addition, we miss one important criterion, which is the possibility to influence the area by active intervention. Actually, we need a comprehensive conceptual model, which makes clear how nature, mobility, and housing tie together, how they interact with other determinants of well-being, and what their impact is on well-being. Such a model was e.g., developed for the Dutch Public Health Status and Forecasts report (Ruwaard & Kramers 1998) and implies the recognition of several groups of determinants of health, including lifestyle, the social and the physical environment. Nature, mobility, and housing could be placed in this scheme. We presume that in the NEHAP context well-being is taken as a widened concept of health.

The second topic concerns targets and indicators. Table 4-1 formulates several targets and areas of intervention. First, the potential effect of the intervention in terms of well-being as an outcome is not given. Secondly, the most concrete part is the indicators. How are they defined and how are they measured? The text refers to a baseline assessment in 1999, but we as readers would have liked to see more details on this. In this context it is noteworthy that the WHO European Centre for Environment and Health recently developed a comprehensive set of environmental health indicators for use in NEHAPs (World Health Organization 2000).

The third topic is concerned with the phases after the initial plan. Under "weaknesses", the authors indicate the lack of a clear implementation strategy along with adequate financing, and the absence of a clear involvement of the general public and the economic sector, whereas these are crucial success factors. Which is then the status of the "interventions" mentioned in Table 4-1? And which is the ex-ante cost-benefit estimate of the plan? In this respect we can learn from the USA. Here, the definition of goals and quantitative targets in the field of environmental health promotion are included in the comprehensive Healthy People Initiative of the Department of Health and Human Services, Washington, DC. In this initiative both the public and the economic sectors are intensively involved in the planning and implementation strategy (U.S. Department of Health and Human Services 2000).

In conclusion, we fully agree that an intersectorial approach is essential in improving our health. We support the initiative of formulating goals and targets, which can be very stimulating. However, in order to be successful, the approach needs to be well thoughtout taking into account all critical phases of the process from monitoring targets to implementing effective interventions.

References

Ruwaard D,Kramers PGN (1998). Public health status and forecasts 1997: health, prevention and health care in the Netherlands until 2015.

Maarssen: Elsevier/De Tijdstroom. U.S. Department of Health and Human Services (2000). Healthy people 2010: understanding and improving health. Washington, DC: DHHS.

World Health Organization (2000). Environ-mental health indicators: development of a methodology for the WHO European Region. Interim Report November 2000. Copenhagen: WHO, Regional Office for Europe.

PART III:

EVALUATION OF ENVIRONMENTAL HEALTH

PROMOTION PROGRAMS

Introduction

The evaluation of the Swiss NEHAP is presented in more detail in chapter 5 at the beginning of this part. The evaluation is carried out at the level of the national program, thus local projects have to be evaluated individually. The comprehensive evaluation (see chapter 2.2) consists of the continuous analysis of the implementation of the program (process evaluation) as well as the assessment of aim-related outcomes and a selected number of more distal impacts (outcome and impact evaluation). In the process evaluation, a descriptive strategy is applied, while a normative approach is used for the evaluation of outcomes and impacts. Based on impact models, a number of specifically adapted indicators have been developed for the evaluation of the Swiss NEHAP.

Meanwhile, the WHO started with the development of a set of environmental health indicators for international application. As a contribution to the ongoing discussion about the different approaches in relation to environmental health indicators and their application, the WHO indicator set will be compared with the Swiss evaluation indicators in the second section of part III (chapter 6). Additionally, the suitability of such environmental health indicators for policy evaluation will be discussed.

5   Evaluation of the Swiss National Action Plan Environment and Health

The evaluation concept for the Swiss NEHAP was developed in 1997,[38] thus in a late phase of the formulation of the NEHAP, and it is carried out since 1998. In the following, the main results of the evaluation are summarised and recent developments due to the evaluation results are described.

5.1   Process evaluation

The implementation process is crucial in environmental health promotion programs. In the evaluation of the Swiss NEHAP this has been taken into account by setting an adequate emphasis on the process evaluation which is carried out continuously.

5.1.1 Summary of the first process evaluation: internal view

A first intermediate report was published in 1999 after the first year of implementation of the NEHAP.[94] It was based on interviews with the project manager from the Swiss Federal Office of Public Health (FOPH), Environment and Health Unit, and all persons involved in the implementation process in various federal offices as well as in the cantons.

Three working groups were built during 1998: one within the FOPH, another with representatives of the other concerned Federal Offices such as Environment, Transport, Agriculture, and Housing, and a third with representatives from the cantons which have their own competencies in Switzerland (e.g. most laws are implemented on a cantonal level). 23 of the 26 cantons had named at least one representative either from the health or the environmental sector as coordinator for the implementation on the cantonal level. It was also planned to name a federal coordinator from the relevant federal office for each intervention area of the NEHAP (see chapter 4.2.2) who should take the lead in the implementation. This succeeded only for 8 of the 14 areas until 1999.

The information flow between FOPH-departments and with the representatives from the federal offices and from the cantons was ensured through regular meetings. While a majority of these representatives identified themselves with their role in the implementation of the NEHAP, this was not the case for the federal intervention area coordinators. They did not hold meetings in 1998 and accordingly, they lacked a common identity within the program. The program management was named as main information source for the NEHAP by all interview partners. A majority of them stated that they had made new contacts through their participation in the NEHAP. Additionally, the interview partners had to state how likely they thought it was that the NEHAP could introduce changes within the next ten years. The range of assessments was quite large with the highest share in the answer-category "maybe" (39%). The cantonal representatives assessed the likelihood slightly more sceptical than the rest of the interview partners. The reservation named most often were the restricted resources of the program. Some of the cantonal representatives also criticised that the implementation was mainly inter-administrational and recommended a stronger political involvement.

5.1.2 Summary of the second process evaluation: external view

In 2000, interviews were carried out with relevant, but not directly involved institutions and interest groups to assess their perception of the NEHAP and its main objectives.[95] The aim of this second series of interview was to provide information allowing an optimisation of the implementation process, to identify additional partners outside the administration, and to point out opponents of the program which could hinder it.

The 27 interview partners were chosen based on an analysis of the relevant societal actor groups in each of the fields addressed by the NEHAP. For example, in the field "indoor environment" the following main actors were identified:

- constructors (represented e.g. by the engineers and architects association),
- house owners/landlords (represented e.g. by the association of house owners),
- users (represented e.g. by the tenants association).

After two years of implementation, about 60% of the interviewed persons did not know the Swiss NEHAP (these persons were provided with written information on the program to allow them nevertheless an assessment of its contents). A very large majority judged the central idea of the Swiss NEHAP, the promotion of health and wellbeing in a healthy environment, as very or rather important (70.4% and 25.9%, respectively). About two thirds of the interviewed persons were in favour of the objectives in the area of "Housing and wellbeing" (see chapter 4.2.2), and half of them supported the objectives in the area "Nature and wellbeing", but it was only about one third in the area of "Mobility and wellbeing". Almost similar proportions of interview partners assessed the likelihood of changes introduced by the NEHAP as high (30%) and low (33%), respectively. Nevertheless, more than three quarters of the interviewed persons wished to be informed on the further development of the program.

5.1.3 Resources for the implementation of the Swiss NEHAP

In the first year, the Swiss NEHAP had to be implemented with quite modest means, disposing of a direct financing by the FOPH of 275'000 CHF and a 70% post, held by the project manager (table 5-1). It has to be noted, however, that many of the topics addressed in the NEHAP are not under the competence of the FOPH and thus financed by other federal offices. Therefore, some of the normal activities of these offices had now also become part of the NEHAP implementation without the explicit allocation of a separate budget. Therefore, it was not possible to compile a complete overview of all financial and personnel resources for the NEHAP.

Table 5-1:   Overview of resources directly allocated to the Swiss NEHAP by the Swiss Federal Office of Public Health 1998-2001.

Year	Financial resources (CHF)	Personnel resources (100% posts)
1998	275'000	0.7
1999	760'000[*]	1.8
2000	900'000[*]	1.8
2001	1'600'000[*]	2.5

[*] *including costs for personnel except project manager*

Even though the resources have increased steadily over the last four years, the overview shows that they are still quite limited. For the following years until 2006, 1.4 Mio. CHF per year are budgeted as direct project funding by the FOPH (personnel costs not included).

5.2   Implementation strategy for the Swiss NEHAP

After three years of process evaluation, a synthesis was compiled in 2000.[96] As a major consequence of this synthesis, the project management developed an implementation program for the years 2001-2006, specifying process targets, the applied strategies, and instruments for the Swiss NEHAP which had been lacking until then. During the development of this implementation program, it became apparent that it would not be possible to reach the aims formulated for the three topics Mobility, Housing, and Nature until 2007[40] on a national level with the resources at hand. Thus, it was determined in the new strategy to limit the aim-related implementation to three pilot regions to identify and disseminate successful approaches and ideas as examples to stimulate similar projects throughout the country later on.[91] In these pilot regions, additional financial and communication means are provided to translate each of the three NEHAP topics into action in an exemplary way. Besides funding of up to 50% of total project costs, the local project teams obtain professional support by the FOPH in the development of their projects. A communication platform is provided as well.

For the national level, objectives were redefined based on the aims of the Swiss NEHAP. These objectives were limited to areas which are under the direct competence of the project management in the Federal Office of Public Health, mainly in the field of information and knowledge transfer. An increased cooperation between public as well as private institutions in the environment and health field was formulated as general objective of the implementation program. Communication and PR, an "innovation pool" to support innovative projects in the 14 intervention fields, and networking were defined as instruments to reach these aims.[91] The binding commitment of partners has now been defined as a long-term objective in the field "networking", thus it is no longer planned to name national coordinators for each of the 14 intervention areas (see chapter 5.1.1). Based on the networking activities, a follow-up NEHAP-program with a wider group of responsible partners shall be set up in 2006. In the meantime, the networking activities shall lead to an integration of NEHAP objectives into activities of partners within the Federal Office of Public Health and in other concerned Federal Offices, as well as in cantons, communities, private organisations, and selected NGOs. It is also planned to carry out common activities within the framework of the NEHAP.

The evaluation concept is currently adapted to the changes introduced by this new implementation strategy. On the one hand, the process evaluation will even gain importance in view of the significance which the implementation process has shown to have for the success of the program. On the other hand, the outcome and impact evaluation will be reoriented focusing on the pilot regions. In the following section, the results of this part of the evaluation as yet carried out will be presented and the adaptations to the new implementation strategy will be outlined.

5.3 Outcome and impact evaluation

5.3.1 Methods and adaptations

Originally, a goal-oriented or "distance to target" approach was applied to evaluate the effectiveness of the NEHAP in reaching its aims. To define appropriate indicators for the Swiss situation, impact models (see chapter 2.2) for each of the three topics were formulated. The hypothesis for the models were based on an extensive document analysis of background material used during the development of the NEHAP, and minutes of the workgroup meetings. Draft versions of the impact models were discussed with members of the working groups to ensure that the models reflected the underlying assumptions of the workgroups and not the views of the evaluation team. These impact models revealed differences in the level of measurability of the aims. While most targets already were operationalized in a measurable way (see chapter 4.2.2 and table 4-1), especially in areas which are politically sensitive (like the impact threshold levels of the Ordinance on Air Pollution Control) or which still lack scientific basis (like the definition of "housing quality", see chapter 3[97]) weaknesses became apparent.

Nevertheless, indicators for the evaluation of outcomes and impacts of the Swiss NEHAP had to be chosen based on these models. An extensive list of 63 indicators was developed, which subsequently was reduced to the 38 most important indicators due to limited data availability and resources. In the tables 10-1, 10-2, and 10-3 (see in the annex), examples from the baseline assessment of these indicators are presented. In chapter 5.3.2, the main results for each of the three topics are summarised.

As already explained in chapter 2, an often encountered problem in the evaluation of environmental health promotion programs is the so called "control group dilemma". Due to this difficulty, a normative instead of a causal approach had been applied originally in the evaluation of the Swiss NEHAP (see p. 68). Currently, the evaluation of outcomes and impacts is adapted to the new implementation program: a limited number of indicators on each topic will be assessed in the respective pilot regions using the same methods as in the national surveys. This approach allows to compare outcomes in the pilot regions, where additional means are provided, with a national "background" and thereby, to make an estimation of the "attributable fraction" of the NEHAP-pilot region projects. Additionally, all pilot regions will carry out local evaluations of process and outcomes of each project which will allow a detailed insight into the project implementation and provide further basis to understand success or failure. This new approach may be a step towards a solution of the "control group dilemma".

5.3.2 Summary of the national baseline assessment

The baseline assessment in the three topics of the NEHAP was carried out 1999/2000 to document the national situation before the start of the program. The assessment was mostly based on data from time series or repeated cross sectional surveys. Importance was attached to the possibility to disaggregate the data in order to identify problem groups (e.g. regions, sex, age, income etc.). In total, 19 different data sources are used for the Swiss evaluation, ranging from the census, micro-censuses on health and traffic or the national monitoring system on air quality to relatively small surveys on housing quality or environmental tobacco smoke.

Agriculture, Nutrition and Wellbeing

In this field, the Swiss NEHAP aims at 75% of the population being in a position to consume healthy and balanced food, including environmental aspects of food consumption like harmony with seasons, regionally produced food and type of production, and thus to contribute to sustainable agriculture (see table 4-1). While over two thirds of the population paid attention to the type of food they consumed (not too much fat, enough vegetables/fruit etc.), only 44.7% considered seasonality when buying food, 34.2% the geographic origin of a product and 24.4% the type of production e.g. organic (table 10-1, annex). Only 51.3% had a good knowledge of seasonality and there existed a discrepancy between knowledge and behaviour. A further target of the NEHAP is the promotion of organic production. In future, 30% of the agriculturally productive land should be cultivated organically whilst in 1998 only 6.7% was cultivated this way.

Mobility and Wellbeing

An important objective in the area "Mobility and Wellbeing" is the attainment of the Swiss Air Quality Standards (table 10-2, annex). The baseline assessment showed that 30.5% of the population were exposed to NO_2-levels above the standard (30 µg/m³) and over 61% to increased PM10-levels (above 20 µg/m³). A further objective of the NEHAP is the doubling of journeys made by bicycle as an ideal form of ecologically not detrimental form of mobility combined with exercise. In 1994, the bicycle was used for 5 to 7% of journeys. Nevertheless, the large proportion of short journeys made by car demonstrates the potential for non-motorized mobility. harmless

Housing and Wellbeing

An objective in the area "Housing and Wellbeing" is the reduction of exposure to involuntary environmental tobacco smoke (ETS) (table 10-3, annex). Baseline data indicates that over 50% of non-smoking Swiss were exposed to ETS at the workplace and 67.8% reported to be annoyed by ETS in restaurants. 44.7% of Swiss schoolchildren were exposed to ETS at home. While a large proportion of the population was satisfied with different characteristics of the housing surroundings, 28.3% reported to be regularly annoyed by traffic noise.

6   Environmental health indicators in policy evaluation[*]

Abstract

In carrying out two projects involving environmental health indicators - a national environmental health programme evaluation and an international environmental health indicator system - in parallel, it became apparent that an international indicator set has limitations regarding the evaluation of a national programme such as the Swiss National Environment and Health Action Plan (NEHAP). The international indicator set proposed by WHO serves the structured description of the underlying cause-effect chains, allows an integrated monitoring of the general environment and health situation and provides valuable international comparisons. However, the relevance of an international indicator set varies in the national context. Moreover, it does not allow the evaluation of a national implementation process, which is highly important in assessing success or failure of an environmental health promotion programme. For a comprehensive evaluation of such a programme, a specific evaluation concept derived from the formulated goals and targets needs to be developed with emphasis on evaluation of the implementation process.

---

[*]   *Published as: Kahlmeier S, Braun-Fahrländer C: Environmental health indicators in policy evaluation. European J Public Health: in press.*

The authors are currently involved in two different projects relating to indicators in the environment and health area. We are responsible for the evaluation of the Swiss National Environment and Health Action Plan (NEHAP).[40] These novel instruments for action in the area of environmental health promotion were developed following recommendations made at the European Ministerial Conferences on Environment and Health.[7, 89] Throughout Europe, around 40 NEHAPs have been presented so far. Switzerland was among the first western European countries to develop such a programme. As a consequence of these political activities, in 2000 the World Health Organization (WHO) started the development of a European environment and health monitoring system,[98-100] and recently proposed a first core set of environmental health indicators.[101] The project aims at establishing a comprehensive system for regular reporting on environment and health within the countries as well as on the WHO European level. The system shall also serve Member States to assess the progress and effectiveness in implementing their NEHAPs.[100] The authors are also in charge of the pilot implementation of this indicator set in Switzerland.

In carrying out these two projects - national evaluation and international indicator system - in parallel, it became apparent that an international indicator set has limitations regarding the evaluation of a national programme such as the Swiss NEHAP. In the following, we point out parallels and differences in the two approaches.

6.1 The WHO's environmental health indicators for the European Region

An 'environmental health indicator' (EHI) is a 'measure which indicates the health outcome due to exposure to an environmental hazard', thus consisting of 'an environmental indicator or a health indicator plus a known environmental-exposure health-effect relationship'.[102] Definitions also emphasize the policy relevance of EHIs: they should relate to aspects that are important to policy makers and amenable to control.[99, 102, 103]

Applying the EHI-methodology, a core set of EHIs was developed by WHO that in its current form comprises indicators on 10 different topics, along with some denominator variables (see table 6-1).[101] As theoretical concept, the 'Driving Forces – Pressure – State – Exposure – Effect – Action framework' (DPSEEA) was used to derive the indicators.[102] This framework supports the structured description of the cause-effect chains between human activities and health outcomes. It also facilitates the identification of possibilities for action on the different levels.

However, the WHO EHI project is also confronted with a number of difficulties. The number of times a pollutant exceeds a threshold level is commonly proposed as an EHI (see table 6-1). If these standards are risk based they contain information on the underlying environment and health relationship. Nevertheless, the percentage of the population exposed to exceeded pollution levels and, for future development, an economic valuation of the health burden would be highly desirable in view of the higher information value for policy makers compared to the percentage of exceeded measurements. A first step in this direction has been made in the WHO indicator set by including, for example, the population exposure to ambient air pollutants or the population annoyance by noise (see table 6-1). While data may be available for air pollution, the required information on the population exposure distribution is often lacking in other fields. Another hindrance is that cause-effect chains between environmental exposures and health effects are often complex and precise measures rare.[4, 102, 104, 105]

Table 6-1:   *Overview of the WHO environment and health indicators (as at May 2002)*[7]

Topic	Core indicators	DPSEEA
Air quality	• Passenger transport demand by mode of transport	driving force
	• Road transport fuel consumption	driving force
	• Emissions of air pollutants	pressure
	• Population-based exposure to air pollutants (urban)	exposure
	• Infant mortality due to respiratory diseases	effect
	• Mortality due to respiratory diseases	effect
	• Mortality due to diseases of the circulatory system	effect
	• Policies to reduce environmental tobacco smoke exposure	action
Radiation	• Incidence of skin cancer	effect
	• Effective environmental monitoring of radiation activity	action
Noise	• Population annoyance by certain sources of noise	effect
	• Sleep disturbance by noise	effect
	• Application of regulations, restrictions and noise abatement measures	action
Housing and settlements	• Living floor area per person	state
	• Population living in substandard housing	exposure
	• Mortality due to external causes in children under 5 years of age	effect
	• Scope and application of building regulations for housing	action
	• Land use and urban planning regulations	action
Traffic accidents	• Mortality from traffic accidents	effect
	• Rate of injuries by traffic accidents	effect
Water and sanitation	• Waste water treatment coverage	pressure
	• Exceedance of recreational water limit values / microbiological parameters	state
	• Exceedance of WHO drinking water guidelines for microbiological parameters	state
	• Exceedance of WHO drinking water guidelines / chemical parameters	state
	• Access to safe drinking water	exposure
	• Access to adequate sanitation	exposure
	• Outbreaks of water-borne diseases	effect
	• Diarrhoea morbidity in children	effect
	• Effective monitoring of recreational water	action
Food safety	• Monitoring chemical hazards in food: potential exposure	exposure
	• Outbreaks of food-borne illness	effect
	• Incidence of food-borne illness	effect
	• General food safety policy	action
	• Effectiveness of food safety controls	action
Waste and contaminated land	• Hazardous waste generation	pressure
	• Contaminated land area	state
	• Hazardous waste policies	action
Chemical emergencies	• Sites containing large quantities of chemicals	pressure
	• Mortality from chemical incidents	effect
	• Regulatory requirements for land-use planning	action
	• Chemical incidents register	action
	• Poison centre service	action
	• Medical treatment guidelines	action
	• Government preparedness	action
Workplace	• Occupational fatality rate	effect
	• Rates of injuries	effect
	• Sickness absence rate	effect
	• Statutory reports of occupational diseases	effect

6.2 The Swiss National Environment and Health Action Plan and its evaluation

The development process of the Swiss NEHAP and its targets have already been discussed in detail[106] (see chapter 4) and therefore will only be presented in brief here: based on an analysis of the Swiss situation, Swiss authorities decided to set priorities in three areas with a need for action in which the association between environment and health can be communicated easily: Mobility and Well-being, Housing and Well-being, and Nature and Well-being (dealing with nutrition and agriculture).[40] The Swiss NEHAP was specifically designed as an environmental health *promotion* programme aiming at complementing already ongoing activities.[106] In each of the three areas, specific and mostly quantified targets were formulated. For example, the fact that in 1994 60% of journeys made by car were no longer than six kilometres demonstrates a large potential for non-motorized mobility in Switzerland. Accordingly, in the area 'Mobility', one target is the doubling of journeys made by bicycle as an ideal form of environmentally friendly mobility combined with exercise.

For the evaluation of the Swiss NEHAP, a comprehensive approach was applied, including planning and implementation as well as outcomes and impacts (see also chapter 5).[35]

In relation to implementation as well as evaluation it is important to remember that health promotion aims not only at the improvement of individual outcomes, but just as much at the change of political, organizational, and social conditions.[14] This is especially true for an environmental health promotion programme like the Swiss NEHAP which is confronted with the difficulty that environment and health departments still operate within largely separated administrative structures in many European countries.[16, 106, 107] Thus, understanding the implementation process of such an intervention (process evaluation) and associated structural changes is of special importance in this field. Such changes in conditions should be seen as 'outcomes' of their own and additionally, they are the basis to understand success or failure in achieving quantified outcomes.[20-22, 108]

Accordingly, emphasis was laid on the process evaluation in the Swiss NEHAP. The mostly qualitative data are collected by repeated interviews with the programme manager and staff as well as the partners involved in the implementation process. A NEHAP-project-database provides information on projects carried out in relation with the NEHAP. Information on the resources available for the implementation, the programme management structure and the ongoing activities (output) are also collected. Important political decisions relating to NEHAP topics are documented to allow a statement on the 'societal climate'. Additionally, a flexible user-focused approach is applied to provide additional information according to the needs of the programme management. As a result of this process evaluation, an implementation strategy for the Swiss NEHAP was developed recently (see chapter 5.2).[109] The implementation will now be focused on three pilot regions and public relations will be intensified.

To define appropriate indicators for the Swiss outcome evaluation (see chapter 5.3), impact models for each of the three topics were formulated. Consisting of hypotheses on the presumed relationship between the programme measures and expected outcomes, they serve as a basis to understand why targets were reached or what impeded programme success.[35] Additionally, potential weaknesses in conceptualization and formulation of targets become apparent. The formulation of such a programme impact theory also facilitates the consideration of intermediate factors not contained in the programme but which might affect goal attainment. For example, in relation to the target of doubling the journeys made by bicycle, not only the share of bicycle traffic should be evaluated, but also intermediate factors such as the access to a bicycle, the availability of a car parking space at the workplace, bicycle facilities at train stations, the development of accidents, or the number of short journeys made by car should be included. In this way, indicators for the Swiss NEHAP evaluation were developed based on the impact models. A baseline assessment of the three topics of the NEHAP was carried out in 1999/2000 to document the situation before the start of the programme, against which progress can be compared later, applying a distance-to-target approach (see chapter 5.3.2).[34]

Figure 6-1: **Development and application of environmental health indicators and indicators for environmental health policy evaluation in Switzerland**

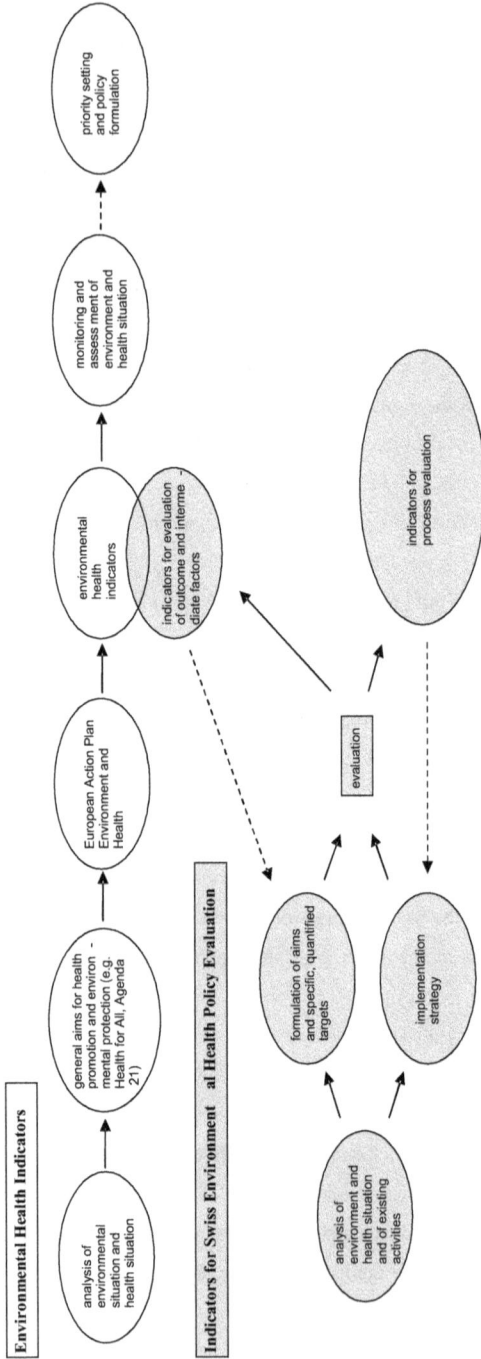

Environmental Health Indicators

analysis of environmental situation and health situation

general aims for health promotion and environmental protection (e.g. Health for All, Agenda 21)

European Action Plan Environment and Health

environmental health indicators

monitoring and assessment of environment and health situation

priority setting and policy formulation

Indicators for Swiss Environmental Health Policy Evaluation

analysis of environment and health situation and of existing activities

formulation of aims and specific, quantified targets

implementation strategy

evaluation

indicators for evaluation of outcome and intermediate factors

indicators for process evaluation

⟶ = steps in the development of indicators

- - -▶ = intended application of indicators

6.3   Parallels and differences between the two approaches

Figure 6-1 illustrates the development process of EHIs compared to indicators for the evaluation of a specific environmental health programme such as a NEHAP. Derived from general analyses of the environment and health situation, the EHI system proposed by WHO covers a wide range of issues, thus allowing integrated monitoring of the general environment and health situation. In countries like Switzerland, which don't have a tradition in environmental health reporting, such a general overview will be particularly useful. Additionally, the currently ongoing pilot implementation of the EHI core set in over a dozen European countries will allow valuable international comparisons.[101]

However, from the point of view of policy evaluation, the relevance of the suggested EHIs varies in the national context. Indicators for the evaluation of a national policy are derived from previously formulated, specific policy targets such as the ones in the Swiss NEHAP (figure 6-1). Therefore, international EHIs are only suitable for the evaluation of a national policy when they coincide with the national priority setting and address areas where action is taken within a country. In this case, national outcome and international EHI sets can partly overlap (figure 6-1), whereby the degree of overlap may vary from country to country. For Switzerland, this is for example the case in the topics of outdoor air quality, noise and traffic accidents (table 6-1). Additional indicators were derived based on the targets formulated in the Swiss NEHAP and the impact models, including intermediate factors. The most important restriction of the WHO indicator set for policy evaluation is, however, that it does not allow the evaluation of a national implementation process, which is highly important in assessing success or failure of an environmental health promotion programme. These indicators have to be derived from and adapted to the respective programme and the national context.

6.4 Conclusions

We conclude that the DPSEEA framework applied by WHO serves the structured description of the cause-effect-chain of known environment and health relationships. An international set of EHIs based on this framework is useful for monitoring purposes as well as international comparison and priority setting. However, its suitability to evaluate progress and effectiveness of the implementation of the Swiss NEHAP is limited. For a comprehensive evaluation of such a programme, a specific evaluation concept derived from the formulated goals and targets needs to be developed with an emphasis on the evaluation of the implementation process.

PART IV:

GENERAL DISCUSSION

7 General discussion and outlook

In the last chapter, the findings of this thesis and implications for the development, implementation and evaluation of environmental health promotion are discussed. A brief outlook on further developments and future activities concludes the chapter.

7.1 Development of environmental health promotion programs

In the Ottawa charter, a number of general principles are outlined which should be applied in health promotion programs[14] (see chapter 2.1). One of those principles is the involvement of the ones affected by a policy into its planning and development. Environmental health promotion is always an intersectorial activity. Therefore, all relevant actors, ideally within as well as outside the administration, should be involved into the development of a program to ensure their collaboration.[89, 110] In chapter 4 was shown that a good inter-administrational involvement was achieved in the development process of the Swiss NEHAP. The formulation of the contents lead to a collaboration between hitherto mostly separated disciplines. However, two important groups were not involved: the economy and the general public. Consequently, the Swiss NEHAP was not well known to persons outside the directly involved circle after two years of implementation (see chapter 5.1.2). However, the non-involvement of the economy is not unique to the Swiss NEHAP: in 1999 the WHO stated that "collaboration with economic sectors has been one of the most difficult areas in the development of NEHAPs in most countries".[89]

Clearly, it is a challenging task to involve "the general public" into the elaboration of a national program. Public hearings are one possibility, but there are also more sophisticated tools such as the "Zukunftswerkstatt", a method recently applied in a project on the development of urban environments in Switzerland.[111] Such approaches, however, are time consuming and the necessary resources need to be provided. Therefore, it was not possible to apply them in the development of the Swiss NEHAP (see chapter 4.3).

Programs for environmental health promotion– as any health promotion strategy – should be evidence based.[17, 110] During the elaboration of the Swiss NEHAP, a gap of knowledge was identified regarding housing quality and wellbeing. The study on perceived housing quality and wellbeing of movers presented in chapter 3 provides insight into this issue: Firstly, a higher satisfaction with environmental housing quality indeed was associated with an improved wellbeing of movers. Secondly, the positive association with environmental indicators was persistent in participants who had moved for other than environmental reasons. Therefore, the results of the study support the view that housing quality and wellbeing is one of the fields were action is justified in urban areas of Switzerland, despite the already high environmental quality. However, several points are of relevance in the discussion of implications of this study for the formulation of measures on this topic. On the one hand, the physical characteristics of the apartment and the apartment related social environment were just as important for the wellbeing of movers as were environmental characteristics; the former, however, are not amenable to political action. On the other hand, we could not entirely clarify which single factors in the residential environment were most influential. Both environmental indicators "perceived air quality" and "location of the building" seemed to reflect a group of different determinants. This result is in accordance with Van Poll's, who also found that not only physical but also other attributes (such as psychosocial or aesthetical) were important for the satisfaction with the residential environment.[63] These findings lead to the conclusion that in projects to improve the housing quality in urban settings, an integrated approach should be applied instead of focusing on single aspects such as "green spaces". It has to be recognized as well that evaluations of the residential environment can differ significantly across different neighbourhoods.[63] A study on the satisfaction with selected housing quality indicators in Switzerland also showed that while the general satisfaction with the residential environment was high[*], differences where found in different regions of the country and across population subgroups.[112] Therefore, a careful assessment of the respective situation is advisable. This is now applied in the NEHAP pilot region on "Housing quality and Wellbeing" (see chapter 5.2), where in parallel with the project development, a survey is carried out to identify main problem areas.

---

[*] *e.g." very" or "rather" satisfied with green spaces in the neighbourhood: 92.6%, with shopping facilities: 84.9%, with the supply with public transportation: 84.7%, with leisure facilities: 76.6%, respectively.*

While the theoretical basis was most limited in the field of housing quality, weaknesses were also identified in the other NEHAP topics. For instance, the impact model for the topic "Nature and Wellbeing" revealed that the main emphasis had been laid on the education of the population, assuming that an increased knowledge would lead to the desired nutritional and consumer behaviours. It is known, however, that information campaigns can hardly influence the nutritional behaviour, since it is also strongly determined by e.g. availability, marketing, price, and personal preferences.[113] Additionally, there was a lack of data on the knowledge and behaviours of the population regarding the association between agriculture, nutrition and health.[38] The impact model on the topic "Mobility and Wellbeing" also showed a strong emphasis on educational aspects, and a lack of data was detected regarding "human powered mobility" (such as walking and cycling).[38]

Despite its limitations, the Swiss NEHAP can still be seen as a positive example for an environmental health promotion program. The baseline assessment has confirmed that there is still a need for action in a number of environmental health domains also in an industrialized country like Switzerland (see chapter 5.3.2). Moreover, the Swiss NEHAP went a step beyond the simple collection of already ongoing activities with the formulation of "ideal situations" (see chapter 4.2.2), goals and objectives (of which most are "SMART" objectives, see p. 26), thus creating a vision for the future.

7.2   Implementation of environmental health promotion programs

While in the Swiss NEHAP, unlike in many others programs,[15] objectives were formulated in a quantified and time bound way, the implementation had not been addressed adequately. In chapter 4.3 it has been concluded, that the implementation strategy should be planned ideally in parallel with the development of the contents of a program. As a precondition, adequate financing is necessary to develop such an implementation strategy and to translate the program into action successfully. In the case of the Swiss NEHAP, it was only after over two years of implementation that the resources for the development of an implementation program where at hand. It revealed that the available resources were insufficient to reach the ambitious goals.

Consequently, the objectives had to be redefined (see chapter 5.2). It has also been recognized that a long term perspective will be necessary to achieve truly intersectorial collaboration and structural changes: On the one hand, the process evaluation has shown that the degree of cooperation of the administrational partners still depended on the individual representatives (see chapter 5.1). The binding commitment of other Federal Offices had to be redefined as a long-term objective since the designation of national coordinators for each of the 14 intervention areas of the NEHAP could not be achieved (see chapter 5.2). On the other hand, only one Federal Office has been assigned with the project management of the NEHAP so far. However, in chapter 4.3 it has been concluded that intersectorial structures are a necessary condition for a sustainable success of national environmental health programs. The lack of such structures on the federal level in Switzerland is still an unsolved issue.

Additionally, the implementation of environmental health promotion should not be merely inter-administrational but all relevant partners should be involved (see chapter 2.1).[89, 110] Since one of the goals of health promotion is to influence the conditions in order to facilitate "healthy choices", the economy is a key player. This was reaffirmed lately by the WHO stating that "unless economic sectors are mobilized as key partners in implementing NEHAPs, the environment and health sectors will make little progress towards their objectives".[89] Additionally, stakeholders as well as cantons and communities are important partners for the local implementation. As described in chapter 5.2, it is now planned to involve these partners more in the implementation of the Swiss NEHAP, while an adequate strategy to involve the economy has yet to be defined.

7.3   Indicators and evaluation of environmental health promotion programs

In chapter 6 it was shown that the suitability of an internationally developed set of environmental health indicators for the evaluation of national environmental health promotion programs such as NEHAPs is limited. For the systematic evaluation of such programs, a specifically adapted set of indicators, derived from operationalized program objectives and based on impact models, has to be defined. Thus, the purpose of indicators systems should be clearly defined (e.g. priority setting or evaluation of a program). Indicator systems usually cannot serve several purposes at a time, since the development processes and consequently, the composition of indicator systems will differ according to their purpose. For instance, the Swiss indicators for sustainable development shall allow a monitoring of the general development of sustainability in Switzerland.[114] However, the indicator system does not refer to the Swiss Actionplan for Sustainable Development[115] and therefore is unsuitable for a comprehensive policy evaluation (implementation and goal attainment).

There are a number of indicator systems which serve international comparisons. These indicator systems, however, usually suffer from the difference in priorities of different regions of the world. E.g. the United Nations Commission of Sustainable Development (UNCSD) presented a set of sustainable development indicators.[116] A recent evaluation of this indicator system in the Swiss context has shown, however, that part of these indicators are of limited relevance for an industrialized country like Switzerland whereas other relevant issues are not covered adequately (e.g. health).[114, 117] The same problem, although to a lesser extent, is encountered in the ongoing WHO project for the development of a core set of environmental health indicators (see chapter 6).[118] Thus, the question arises as to the feasibility of one common set of indicators for international or even global comparisons.[119] In a globalised world, benchmarking is of increasing importance. However, the considerable differences in priorities and preconditions may make such a task very challenging,[105] and, from a certain point of view, even not desirable. Existing indicator sets for international comparisons usually are more adapted to the priorities and needs of less developed countries. Thus, in industrialized countries they might lead to the - erroneous - conclusion that there is no need for action in the field, while a set of indicators which is specifically adapted to the national priorities might prove the opposite (see section 7.1 above). Thus, a set of indicators which allows international benchmarking and addresses national priorities at the same time remains to be defined. In the WHO environmental health indicators project, it is planned to address this problem by complementing the core set with additional indicators according to national priorities.[100]

Due to the importance of the implementation process in environmental health promotion programs, a strong emphasis needs to be laid on the process evaluation. As explained in chapter 6, the evaluation of the implementation process is crucial, since it allows a continuous improvement of the implementation. Moreover, it improves the understanding of success or failure in achieving outcomes.[20-22] Additionally, achievements which can be linked directly to the program can be identified such as new intersectorial structures, increased collaboration, or learning processes. Therefore, the completion of the WHO environmental health (outcome) indicator set with a number of process indicators relating to e.g. administrative structures, resources for the implementation of environmental health policies within the countries, or to intersectorial decision making mechanisms[108] would be useful.

7.4 Outlook

This dissertation was carried out in a field of increasing relevance and consequently, in parallel with a number of ongoing activities. Lessons and experiences from this thesis enter already into the discussion in ongoing projects in Switzerland such as the monitoring of sustainable development (see chapter 7.3 above) or a planned health monitoring system ("observatoire de santé"). The author is also involved in the European WHO project on environmental health indicators (see chapter 6 and 7.3). Accordingly, a number of new topics emerged in the course of the work.

The questions remains as to which would be an appropriate set of environmental health indicators for an international comparison, that at the same time responds to national needs. One possibility, which will be explored in the WHO indicator project, is the application of "reference values" which may vary in different European regions while at the same time, comparability of the indicators would be maintained.[120] A (preliminary) final set of core indicators will be defined by the end of this year and the assessment in a number of pilot countries will start in early 2002. Based on the results of this project, it is planned to present a first evaluation of the environmental health situation in Europe at the 4th Ministerial Conference on Environment and Health in 2004 in Budapest. Another challenge for the future lies in the definition of health indicators in the framework of sustainable development monitoring since the health indicators proposed by UNCSD are not appropriate for industrialized countries like Switzerland.

The evaluation of the Swiss NEHAP will also continue. In 2004, an intermediate evaluation report will be presented which will mainly focus on results from the process evaluation with a special emphasis on the pilot regions (see chapter 5.3.1). Hopefully, it will allow a more detailed insight into successful strategies in environmental health promotion.

PART V:

INDEXES AND ANNEX

8 References

1. WHO Regional Office for Europe, *Environment and Health: The European Charter and Commentary*. European Series. Vol. 35. 1989, Copenhagen: World Health Organization.

2. Smith, K.R., C.F. Corvalan, and T. Kjellstrom, *How much global ill health is attributable to environmental factors?* Epidemiology, 1999. **10**(5):573-84.

3. Corvalan, C. and T. Kjellstrom, *Health and environment analysis for decision making*. World Health Stat Q, 1995. **48**(2):71-7.

4. Cole, D.C., J. Eyles, and B.L. Gibson, *Indicators of human health in ecosystems: what do we measure?* Sci Total Environ, 1998. **224**(1-3):201-13.

5. WHO Commission on Health and Environment, *Our planet, our health*. 1992, Geneva: World Health Organization.

6. WHO Regional Office for Europe, *Targets for Health for all: the health policy for Europe*. European Health for All Series, ed. WHO. Vol. 4. 1993, Copenhagen: WHO.

7. WHO Regional Office for Europe. *Environmental Health Action Plan for Europe*. in *Second European Conference on Environment and Health*. 1994. Helsinki, Finland: World Health Organization.

8. WHO Regional Office for Europe. *Overview of the environment and health in Europe in the 1990s: Background document*. in *Third Ministerial Conference on Environment and Health*. 1999. London.

9. WHO Expert Committee on Environmental Health in Urban Development, *Environmental health in urban development: report of a WHO expert committee*. reprinted 1995 ed. technical report series. Vol. 807. 1991, Geneva: World Health Organization.

10. Malmstrom, M., J. Sundquist, and S.E. Johansson, *Neighborhood environment and self-reported health status: a multilevel analysis*. Am J Public Health, 1999. **89**(8):1181-6.

11. Yen, I.H. and G.A. Kaplan, *Poverty area residence and changes in depression and perceived health status: evidence from the Alameda County Study*. Int J Epidemiol, 1999. **28**(1):90-4.

12. Yen, I.H. and G.A. Kaplan, *Neighborhood social environment and risk of death: multilevel evidence from the Alameda County Study*. Am J Epidemiol, 1999. **149**(10):898-907.

13. Bertonllini, R., et al., *Environment and Health 1: Overview and main european issues*. WHO regional publications, European series No. 68. 1996, Copenhagen: WHO.

14. WHO Regional Office for Europe. *Ottawa Charter for Health Promotion*. in *First International Conference on Health Promotion*. 1986. Ottawa, Canada: World Health Organization.

15. von Schirnding, Y., *Incorporating health-and-environment considerations into sustainable development planning. Report on a WHO/UNDP Initiative*. 1999, Geneva: WHO Sustainable Development and Healthy Environments.

16. Ziglio, E., S. Hagard, and J. Griffiths, *Health promotion development in Europe: achievements and challenges*. Health Promot Internation, 2000. **15**(2):143-54.

17. Rosenbrock, R., *Public Health als soziale Innovation [Public health as a social innovation]*. Gesundheitswesen, 1995. **57**(3):140-4.

18. WHO, *Health programme evaluation: guiding principles for its application in the managerial process for national health development*. Health for All Series, ed. WHO. Vol. 6. 1981, Geneva.

19. Bussmann, W., U. Klöti, and P. Knoepfel, eds. *Einführung in die Politikevaluation [Introduction in policy evaluation]*. 1997, Helbing & Lichtenhahn: Basel, Frankfurt a.M.

20. Speller, V., A. Learmonth, and D. Harrison, *The search for evidence of effective health promotion*. Bmj, 1997. **315**(7104):361-3.

21. Nutbeam, D., *Evaluating health promotion - progress, problems and solutions*. Health Promot Internation., 1998. **13**(1):27-44.

22. Koelen, M.A., L. Vaandrager, and C. Colomer, *Health promotion research: dilemmas and challenges*. J Epidemiol Community Health, 2001. **55**(4):257-62.

23. WHO, *Preamble to the Constitution of the World Health Organization as adopted by the International Health Conference, New York, 19-22 June, 1946; entered into force on 7 April 1948.* 1948, WHO: New York.

24. World Health Organization. *Sundsvall statement on supportive environments for health.* in *Third International Conference on Health Promotion.* 1991. Sundsvall, Sweden.

25. Black, H., *Environmental and public health: pulling the pieces together.* Environ Health Perspect, 2000. **108**(11):A512-5.

26. Baker, G., *An essay concering the cause of the endemial colic of Devonshire. Read in the Theater of the College of Physicians in London.* 1767, J Hughs near Lincoln's-Inn-Fields: London.

27. Cameron, D. and I.G. Jones, *John Snow, the broad street pump and modern epidemiology.* Int J Epidemiol, 1983. **12**(4):393-6.

28. Hill, A.B., *The environment and disease: Association or causation?* Proc. R. Soc. Med, 1965. **58**:295-300.

29. Rothman, K.J., *Modern Epidemiology.* 1986, Boston/Toronto: Little, Brown and Company.

30. Gochfeld, M. and B.D. Goldstein, *Lessons in environmental health in the twentieth century.* Annu Rev Public Health, 1999. **20**:35-53.

31. Hilleboe, H.E. and A.R. Jacobson, *Environmental health-a conceptual approach.* Arch Environ Health, 1966. **12**(6):787-92.

32. WHO Regional Office for Europe, *Target for Health For All: targets in support of the European Regional Strategy for Health for All.* 1985, Copenhagen: World Health Organization.

33. Keating, M., *Agenda for a sustainable development in the 21st century.* 1993, Geneva: Center for our common future.

34. Bircher, U., et al., *Evaluation des Aktionsplans Umwelt und Gesundheit: Ausgangslage. Kurzfassung (Evaluation of the Swiss National Environment and Health Action Plan: Baseline assessment. Executive summary).* 2000. (http://www.unibas.ch/ispmbs/apug/apughome.htm), Basel: Institute of Social and Preventive Medicine of the University of Basel, Department Environment and Health.

35. Rossi, P. and H. Freeman, *Evaluation: a systematic approach.* 5th ed. 1993, Newbury Park, California: Sage Publications.

36. Swiss Federal Office of Public Health Evaluation Unit, *Leitfaden für die Planung von Projekt- und Programmevaluation [Guidelines for the planning of project and program evaluation].* 1997, Bern: Swiss Federal Office of Public Health.

37. Kunzli, N., et al., *Public-health impact of outdoor and traffic-related air pollution: a European assessment.* Lancet, 2000. **356**(9232):795-801.

38. Kahlmeier, S., et al., *Aktionsplan Umwelt und Gesundheit: Evaluationskonzept [Swiss National Environment and Health Action Plan: Evaluation concept].* 1998, Basel: Institute of Social and Preventive Medicine of the University of Basel.

39. Bussmann, W., *Evaluationen staatlicher Massnahmen erfolgreich begleiten und nutzen [How to accompany and use evaluations of governmental measures successfully].* 1995, Chur/Zürich: Rüegger.

40. Swiss Federal Office of Public Health and Swiss Agency for the Environment Forests and Landscape, *Aktionsplan Umwelt und Gesundheit APUG (Actionplan Environment and Health).* New edition 2001 ed. 1997, Bern.

41. Strachan, D.P. and C.H. Sanders, *Damp housing and childhood asthma; respiratory effects of indoor air temperature and relative humidity.* J Epidemiol Community Health, 1989. **43**(1):7-14.

42. Verhoeff, A.P., et al., *Damp housing and childhood respiratory symptoms: the role of sensitization to dust mites and molds.* Am J Epidemiol, 1995. **141**(2):103-10.

43. Haan, M., G.A. Kaplan, and T. Camacho, *Poverty and health. Prospective evidence from the Alameda County Study.* Am J Epidemiol, 1987. **125**(6):989-98.

44. Mackenbach, J.P., et al., *The determinants of excellent health: different from the determinants of ill-health?* Int J Epidemiol, 1994. **23**(6):1273-81.

45. Krause, N., *Neighborhood deterioration and self-rated health in later life.* Psychol Aging, 1996. **11**(2):342-52.

46.   Lawton, M.P., E.M. Brody, and P. Turner-Massey, *The relationships of environmental factors to changes in well-being.* Gerontologist, 1978. **18**(2):133-7.

47.   Carp, F.M., *Impact of improved housing on morale and life satisfaction.* Gerontologist, 1975. **15**(6):511-5.

48.   Carp, F.M., *Impact of improved living environment on health and life expectancy.* Gerontologist, 1977. **17**(3):242-9.

49.   Krause, N.M. and G.M. Jay, *What do global self-rated health items measure?* Med Care, 1994. **32**(9):930-42.

50.   Kaplan, G.A., et al., *Perceived health status and morbidity and mortality: evidence from the Kuopio ischaemic heart disease risk factor study.* Int J Epidemiol, 1996. **25**(2):259-65.

51.   Idler, E.L. and Y. Benyamini, *Self-rated health and mortality: a review of twenty-seven community studies.* J Health Soc Behav, 1997. **38**(1):21-37.

52.   Miilunpalo, S., et al., *Self-rated health status as a health measure: the predictive value of self-reported health status on the use of physician services and on mortality in the working-age population.* J Clin Epidemiol, 1997. **50**(5):517-28.

53.   Rodin, J. and G. McAvay, *Determinants of change in perceived health in a longitudinal study of older adults.* J Gerontol, 1992. **47**(6):373-84.

54.   Smith, A.M., J.M. Shelley, and L. Dennerstein, *Self-rated health: biological continuum or social discontinuity?* Soc Sci Med, 1994. **39**(1):77-83.

55.   Bobak, M., et al., *Socioeconomic factors, perceived control and self-reported health in Russia. A cross-sectional survey.* Soc Sci Med, 1998. **47**(2):269-79.

56.   WHO Regional Office for Europe, *Twenty steps for developing a Healthy Cities project.* 2nd ed. 1995, Copenhagen: World Health Organization.

57.   Baumgartner, F., *Attraktive und konkurrenzfähige Städte: Kernstadt und Agglomeration - Probleme und Aufgaben einer schweizerischen Agglomerationspolitik.* 2000, Bern: Bundesamt für Raumplanung.

58.   Statistisches Amt der Stadt Bern. Link Institut, *Stadt Bern: Einwohnerbefragung 1996: Fragebogen.* 1996, Bern.

59.   Neue Zürcher Zeitung, *Immobarometer: Fragebogen. Link Institut im Auftrag der Neuen Zürcher Zeitung.* 1998, Luzern.

60.   Bundesamt für Statistik, *Schweizer Gesundheitsbefragung 1997: Telefonischer und schriftlicher Fragebogen.* 1998, Neuchâtel: Bundesamt für Statistik.

61.   Kleinbaum, D.G., L.L. Kupper, and K.E. Muller, *Variable Reduction and Factor Analysis*, in *Applied Regression Analysis and Other Multivariable Methods*, P. Michael, Editor. 1988, PWS-Kent Publishing Company: Boston. p. 595-641.

62.   SPSS Inc., *SYSTAT 7.0 for Windows.* 1997, SPSS Inc.: Chicago.

63.   van Poll, R., *The perceived quality of the urban residential environment: A multi-attribute evaluation [dissertation].* 1997, Rijksuniversiteit Groningen: Groningen.

64.   Röösli, M., et al., *Spatial variability of different fractions of particulate matter within an urban environment and between urban and rural sites.* J Air & Waste Manage Assoc, 2000. **50**:1115-24.

65.   Lufthygieneamt beider Basel, *Die Luftbelastung in der Region Basel: Jahresbericht 1999.* 2000, Liestal.

66.   Bundesamt für Statistik, ed. *Statistisches Jahrbuch der Schweiz 1998.* 1997, Neue Zürcher Zeitung: Zürich.

67.   Imhof, M., *Migration und Stadtentwicklung: Aktualgeographische Untersuchungen in den Basler Quartieren Iselin und Matthäus.* Basler Beiträge zur Geographie. Vol. 45. 1998, Basel: Wepf.

68.   Bentham, G., *Migration and morbidity: implications for geographical studies of disease.* Soc Sci Med, 1988. **26**(1):49-54.

69.   Dalgard, O.S. and K. Tambs, *Urban environment and mental health. A longitudinal study.* Br J Psychiatry, 1997. **171**:530-6.

70.   Marans, R.W., *Perceived quality of residential environment*, in *Perceiving environmental quality : research and applications*, K. Craik and E. Zube, Editors. 1976, Plenum Press: New York. p. 123-148.

71.   Francescato, G., S. Weidemann, and A.J. R., *Evaluating the built environment from the user's point of view: an attitudinal model of residential satisfaction*, in *Building evaluation*, W.F.E. Preiser, Editor. 1989, Plenum Press: New York. p. 181-198.

72.   Goldstein, M.S., J.M. Siegel, and R. Boyer, *Predicting changes in perceived health status.* Am J Public Health, 1984. **74**(6):611-4.

73.   Markides, K.S. and D.J. Lee, *Predictors of well-being and functioning in older Mexican Americans and Anglos: an eight-year follow-up.* J Gerontol, 1990. **45**(2):S69-73.

74.   Mackenbach, J.P., H. van de Mheen, and K. Stronks, *A prospective cohort study investigating the explanation of socio- economic inequalities in health in The Netherlands.* Soc Sci Med, 1994. **38**(2):299-308.

75.   Francescato, G., S. Weidemann, and J.R. Anderson, *Residential satisfaction: Its uses and limitations in housing research*, in *Housing and neighborhoods: Theoretical and empirical contributions*, W. van Vliet, et al., Editors. 1987, Greenwood Press: Westport. p. 43-57.

76.   Evans, G.W. and S. Cohen, *Environmental stress*, in *Handbook of environmental psychology*, D. Stokols and I. Altmann, Editors. 1987, John Wiley & Sons: New York, Chichester, Brisbane, Toronto, Singapore. p. 571-610.

77.   Cassell, E.J., *The right to a clean environment.* Arch Environ Health, 1969. **18**(5):839-43.

78.   Ackermann-Liebrich, U., C. Braun, and R. Rapp, *Epidemiological analysis of an environmental disaster: the Schweizerhalle experience.* Environ Res, 1992. **58**(1):1-14.

79.   World Commission on Health and Environment, *Our common future.* 1987, Oxford: Oxford University Press. 4-5.

80.   Godlee, F. and A. Walker, *Importance of a healthy environment.* Bmj, 1991. **303**(6810):1124-6.

81.   Haralanova, M., *Implementation of National Environmental Health Action Plans in the Czech Republic, Estonia, Lithunia, Poland and the Slovak Republic.* European EpiMarker, 2000. **4**(2):2-6.

82.   Isac, A., *Environment and Health co-operate in the Republic of Moldova.* european bulletin of environment and health, 2001. **8**(1):2.

83.   Kahlmeier, S., N. Künzli, and C. Braun-Fahrländer, *Aktionsplan Umwelt und Gesundheit: Ein wichtiger Beitrag zur Gesundheitsförderung und Prävention in der Schweiz [Action Plan Environment and Health: An important contribution to health promotion and prevention in Switzerland].* Schweiz Ärztezeitung, 1997. **78**(49):1852-54.

84.   Federal Ministry of Environment Youth and Family Affairs, Federal Ministry of Labour Health and Social Affairs, and Bundesministerin für Frauenangelegenheiten und Verbraucherschutz, *Austrian National Environmental Health Action Plan.* 1999, Vienna: Federal Ministry of Environment, Youth and Family Affairs.

85.   van Herten, L.M. and H.P. van de Water, *New global Health for All targets.* Bmj, 1999. **319**(7211):700-3.

86.   Pinter, A., ed. *Hungarian Environmental Health Action Plan.* 1997: Budapest.

87.   Ministry of Health and Ministry of Environment, *National Environmental Health Action Plan for Poland.* 1999, Warsaw.

88.   Ministry of Health Care of Ukraine and Ministry of Environmental Protection and Nuclear Safety of Ukraine, *National Environmental Health Action Plan of Ukraine 1999-2005.* 1999, Kiev.

89.   WHO Regional Office for Europe. *Implementing national environmental health action plans in partnership.* in *Third Ministerial Conference on Environment and Health.* 1999. London, United Kingdom.

90.   Republic of Bulgaria Council of Ministers, *National Environmental·Health Action Plan.* 1998, Sofia.

91.   Environment and Health Unit of the Swiss Federal Office of Public Health,
       *Umsetzungsprogramm Aktionsplan Umwelt und Gesundheit: Ziele, Strategie und Instrumente
       (Schlussentwurf) [Implementation program for the Swiss NEHAP: aims, strategy and
       instruments (final draft)]*. 2001, Bern: Swiss Federal Office of Public Health.

92.   Chartered Institute for Environmental Health, *Integrating environment and health policy: a
       "joined-up" response to recent government consultations.* 1998.
       (http://www.cieh.org.uk/about/policy/integrat/index.htm, accessed 12 March 2001).

93.   World Health Organization, *The Healthy Route to a Sustainable World. Health, Environment
       and Sustainable Development.* 1995, Geneva, Switzerland: World Health Organization, United
       Nations Development Programme.

94.   Kahlmeier, S., et al., *Aktionsplan Umwelt und Gesundheit: Prozessevaluation. Erster
       Zwischenbericht [Swiss National Environment and Health Action Plan: process evaluation. First
       intermediate report].* 1999, Basel: Institute of Social and Preventive Medicine of the University
       of Basel.

95.   Kahlmeier, S. and C. Braun-Fahrländer, *Aktionsplan Umwelt und Gesundheit:
       Prozessevaluation. Zweiter Zwischenbericht: Externe Sicht [Swiss National Environment and
       Health Action Plan: process evaluation. Second intermediate report: external view].* 2000,
       Basel: Institut für Sozial- und Präventivmedizin der Universität Basel.

96.   Kahlmeier, S. and C. Braun-Fahrländer, *Aktionsplan Umwelt und Gesundheit: Erste Synthese
       aus Sicht der Prozessevaluation. [Swiss National Environment and Health Action Plan: first
       synthesis from the point of view of the process evaluation].* 2000, Basel: Institut für Sozial- und
       Präventivmedizin der Universität Basel.

97.   Kahlmeier, S., et al., *Perceived housing quality and well being of movers in Switzerland.* J
       Epidemiol Community Health, 2001. **55**:708-715.

98.   Wills, J.T. and D.J. Briggs, *Developing indicators for environment and health.* World Health
       Stat Q, 1995. **48**(2):155-63.

99.   Briggs, D., *Environmental Health Indicators: Framework and Methodologies.* Protection of the
       Human Environment: Occupational and Environmental Health Series, ed. WHO. Vol.
       WHO/SDE/OEH/99.10. 1999, Geneva: WHO Sustainable Development and Healthy
       Environments.

100.   WHO European Centre for Environment and Health, *Environmental Health Indicators:
       Development of a methodology for the WHO European Region.* 2000, Bilthoven: WHO
       European Centre for Environment and Health (Bilthoven Division).

101.   WHO European Centre for Environment and Health, *Environmental Health Indicators for the
       WHO European Region: Update of Methodology.* 2002: Copenhagen.

102.   Briggs, D., C. Corvalan, and M. Nurminen, eds. *Linkage methods for environment and health
       analysis: General guidelines.* 1996, WHO Office of Global and Integrated Environmental
       Health: Geneva.

103.   Corvalan, C., D. Briggs, and T. Kjellström, *The need for information: environmental health
       indicators,* in *Decision-Making in Environmental Health: from evidence to action,* C. Corvalan,
       D. Briggs, and G. Zielhuis, Editors. 2000, E & FN Spon on behalf of the WHO: London & New
       York.

104.   Pastides, H., *An epidemiological perspective on environmental health indicators.* World Health
       Stat Q, 1995. **48**(2):140-3.

105.   Briggs, D., *Methods for building environmental health indicators,* in *Decision-Making in
       Environmental Health: from evidence to action,* C. Corvalan, D. Briggs, and G. Zielhuis,
       Editors. 2000, E & FN Spon on behalf of the WHO: London & New York. p. 57-75.

106.   Kahlmeier, S., N. Künzli, and C. Braun-Fahrländer, *The first years of implementation of the
       Swiss National Environment and Health Action Plan (NEHAP): lessons for environmental health
       promotion.* Soz Praventivmed, 2002. **47**(2):67-79.

107.   Ritsatakis, A., *Experience in setting targets for health in Europe.* Eur J Public Health, 2000. **10**(4
       Suppl):7-10.

108.   Ziglio, E., *Indicators of health promotion policy: directions for research,* in *Health promotion
       research: towards a new social epidemiology,* B. Badura and I. Kickbush, Editors. 1991, WHO
       Regional Office for Europe: Copenhagen. p. 55-83.

109. Swiss Federal Office of Public Health Environment and Health Unit, *Aktionsprogramm Umwelt und Gesundheit: Ziele, Strategien und Instrumente (Action program environment and health: aims, strategies and instruments)*. 2002, Bern: Swiss Federal Office of Public Health.

110. Corvalan, C., F. Barten, and G. Zielhuis, *Requirements for successful environmental health decision-making*, in *Decision-Making in Environmental Health: from evidence to action*, C. Corvalan, D. Briggs, and G. Zielhuis, Editors. 2000, E & FN Spon on behalf of the WHO: London & New York. p. 11-24.

111. Hodel, J., I. Rihm, and D. Wiener, *Aktionsplan Stadtentwicklung Basel: Feinkonzept [Actionplan for the development of the urban environment in Basel: Detailed concept]*. 1997, Basel: Ökomedia.

112. Haller, D., S. Kahlmeier, and C. Braun-Fahrländer, *Grundlagen zur Evaluation des Aktionsplans Umwelt und Gesundheit: Teilbereich Wohnen und Wohlbefinden. Zufriedenheit mit der Wohnqualität und Freizeitmobilität [Background report for the evalution of the Swiss National Environment and Health Action Plan: Topic "Housing and Wellbeing". Satisfaction with the housing quality and leisure time mobility]*. 2000, Basel: Institute of Social and Preventive Medicine of the University of Basel.

113. Swiss Federal Office of Public Health, ed. *Vierter Schweizerischer Ernährungsbericht [Fourth Swiss Nutrition Report]*. 1998: Bern.

114. Swiss Agency for the Environment Forests and Landscape and Swiss Federal Office of Statistics, *Projekt MONET (Monitoring der nachhaltigen Entwicklung): Projektbeschrieb und Arbeitsplanung. [Project MONET (Monitoring of sustainable development: project description and scheme of work]*. 2000, Neuchâtel.

115. Conseil du développement durable, *Nachhaltige Entwicklung: Aktionsplan für die Schweiz [Sustainable development: action plan for Switzerland]*. 1997, Bundesamt für Umwelt,Wald und Landschaft: Bern.

116. United Nations, *Indicators of Sustainable Development - Framework and Methodologies*. 1996, United Nations: New York.

117. de Montmollin, A. and D. Altweg, *Nachhaltige Entwicklung in der Schweiz: Materialien für ein Indikatorensystem [Sustainable development in Switzerland: materials for an indicator system]*. Statistik der Schweiz, Fachbereich 2 (Raum und Umwelt), ed. Bundesamt für Statistik and Bundesamt für Umwelt Wald und Landschaft. 1999, Neuchâtel: BFS.

118. Haller, D., S. Kahlmeier, and C. Braun-Fahrländer, *WHO Environmental Health Indicators Pilot Project: Feasibility Study. Summary Report from Switzerland*. 2001. (http://www.unibas.ch/ispmbs/pdf/envh_ind.pdf), Basel: Institute of Social and Preventive Medicine of the University of Basel. On behalf of the Swiss Federal Office of Public Health, Health and Environment Unit.

119. WHO Office of Global and Integrated Environmental Health, *Draft Environmental Health Criteria Document: Indicators for decision-making in environmental health*. 1997, Geneva: WHO.

120. WHO European Centre for Environment and Health Bonn Office, *Environmental health indicators pilot project: WHO progress review meeting (draft 24 July 2001)*. 2001, Bonn: WHO.

9   Abbreviations and glossary

95% CI	95% confidence interval (range of values which includes the true value with 95% confidence)
CO_2	Carbon dioxide
DPSEEA	Driving forces – pressure – state – exposure – effect – action framework
ed./eds.	editor/s (or edition)
e.g.	exempli gratia (Latin = for example)
EHI	Environmental health indicator
ETS	Environmental tobacco smoke
FOPH	Swiss Federal Office of Public Health
i.e.	id est (Latin = that is to say)
NGO	Nongovernmental organisation
NEHAP	National Environment and Health Action Plan
NO_2	Nitrogen dioxide
OECD	Organisation for Economic Co-operation and Development
OR	Odds ratio (relative measure of the occurrence of a particular event: ratio of the odds in favour of an event in an exposed group to the odds in favour of the same event in an unexposed group)[132]
PM_{10}	Particles with an aerodynamic diameter of less than 10 μm
SRH	Self rated health
UNCSD	United Nations Commission on Sustainable Development
WHO	World Health Organization

10   Annex

Table 10-1: Targets, examples of evaluation indicators and variables and baseline assessment of the Swiss National Action Plan Environment and Health. Area "Nature and Wellbeing".

Targets	Agriculture, Nutrition and Wellbeing — examples of indicators and variables	Baseline assessment %	Year
• By 2002, 80% of the population will know how to eat healthily and in harmony with seasons and that their consumption patterns influence agricultural production and landscape.	• knowledge on healthy nutrition* - pay attention to something in their nutrition in general • knowledge on seasonality† - good knowledge on seasonality‡ • consideration of environmental criteria when buying food (always/often)† - seasonality - geographic origin - type of production (e.g. organic)	69.0 51.3 44.7 34.2 24.4	1997 1998 1998
• By 2007, nearly 100% of the agricultural soils will be used according to the principles of integrated pest management or organic production (OP), proportion of OP > 30%.	• share of production types on agricultural land§ - organic production - integrated pest management - conventional production	6.7 77.0 16.3	1998
• By 2007, 70% of the available meat will be from species appropriate and livestock-friendly production.	not evaluated		
• By 2007, nitrate content of 99% of all drinking water will be <40 mg/l.	• percentage of drinking water reservoirs with nitrate levels <40mg/l	n.a.	
• By 2007, 90% of all agricultural and related business will have standardised quality control systems; positive declaration / reproducible production pathways the rule.	• share of agricultural businesses with a standardised control system§ - organic production - integrated pest management	7.1 73.9	1998

*Swiss Health Survey (n=13'004), Swiss Federal Office of Statistics †Survey on agriculture, nutrition and health (n=623), Institute of Social and Preventive Medicine of the University of Basel ‡at least seven (of ten possible) correct answers concerning the local season of 5 fruits and 5 vegetables §agrarian information system AGIS (95% of all farms), Federal Office of Agriculture n.a.= data not available yet, monitoring system is established

Table 10-2: Targets, examples of evaluation indicators and variables and baseline assessment of the Swiss National Action Plan Environment and Health. Area "Mobility and Wellbeing".

Mobility and Wellbeing			
Targets	examples of indicators and variables	Baseline assessment %	Year
• By 2002, 80% of the population will know about the interactions between motorised traffic, emissions and adverse impacts on human health.	• knowledge on the association between motorized traffic and health*		
	- share who believes that air quality can influence health	90.0	1999
	- share who believes that noise can influence health	71.0	
• Emissions of motorised traffic will be reduced to such an extent that the impact threshold levels of the Ordinance on Air Pollution Control can be respected.	• share of population who is exposed to air quality levels above threshold		
	- PM10: > 20 $\mu g/m^3$ (annual mean)[†]	61.3	1997
	- NO2: > 30 $\mu g/m^3$ (annual mean)[‡]	30.5	1995
	• modal split of goods traffic crossing the alpine arc[§]		
	- modal split in Mio. tonnes of goods: by railway	72.0	1998
	by lorry	28.0	
• By 2007, the proportion of journeys by bicycle will have doubled for commuting, shopping and leisure as compared to 1995.	• proportion of journeys made by bicycle[¶]		
	- for commuting	7.0	1994
	- for shopping	5.0	
	- for leisure time	7.0	
	• proportion of short journeys made by car[¶]		
	- up to 1 km	10.0	
	- up to 3 km	31.0	
	- up to 6 km	60.0	
	vehicle stock[¶]		
	- percentage of persons that can dispose of a bicycle any time	68.0	
	- percentage of persons that can dispose of a car any time	57.0	
	• security of cyclists**		
	- percentage of totally injured persons by traffic accidents	12.9	1996
	- percentage of totally killed persons by traffic accidents	9.0	

* Eurobarometer (n=1063), Service suisse d'information et d'archivage de données pour les sciences sociales (SIDOS) [†]Health costs due to traffic-related air pollution: PM10 population exposure, Federal Ministry for Environment, Traffic, Energy and Communication [‡]National monitoring system of air quality. Federal Office of Environment, Forests and Landscape [§]Alpinfo (based on traffic counts), Federal Office for Landscape Development [¶]Micro census Traffic (n=18'020), Bureau for Transport Studies **Statistic on Traffic Accidents (all accidents reported to the police), Federal Office of Statistics

Table 10-3: Targets, examples of evaluation indicators and variables and baseline assessment of the Swiss National Action Plan Environment and Health. Area "Housing and Wellbeing".

Targets	Housing and Wellbeing — examples of indicators and variables	Baseline assessment	Year
• By 2002, 80% of the population will be well informed about indoor air pollution and able to take adequate measures.	not evaluated		
• By 2002, a speed limit of 30 km/h will be introduced in 70% of urban and peri-urban residential areas	• development of zones with speed limit of 30km/h[*]		1997
	– nr. of granted zones	356	
	– nr. of maximum possible zones	n.a.	
• By 2002, no-one will be submitted to involuntary passive smoking in the workplace, means of public transport and public buildings.	• exposure to passive smoke at the workplace[†]		
	– % of exposed non-smokers (sometimes/often/always)	51.8	1998
	• exposure of children to passive smoke[‡]		1997
	– % of schoolchildren which are regularly exposed	44.7	
	• annoyance by passive smoke[§] (% of non-smokers, often/sometimes)		
	– at the workplace	30.4	1997
	– in restaurants	67.8	
• By 2007, residential areas will have structures to encourage active involvement in neighbourhood life. Planning interventions will create conditions allowing adequate presence of small manufacturers, jobs (esp. supply), leisure and services.	• satisfaction with characteristics of the housing quality[¶] (% very or rather satisfied)		1998
	– green spaces	92.4	
	– child friendliness	85.9	
	– shopping possibilities	84.1	
	– accessibility by public transport	82.2	
	– leisure facilities	70.7	
	• annoyance at home from external sources[**] (% annoyed regularly)		1997
	– by noise from traffic	28.3	
	– by noise from neighbours	19.3	

[*]VERSIDAT database (based on interviews with cantons), Swiss Traffic Club [†]TRAM-study (n=1201), Institute of Social and Preventive Medicine of the University of Bern [‡]SCARPOL-study (n=4470), Institute of Social and Preventive Medicine of the University of Basel [§]Study on Passive Smoking (n=700), Association for Tobacco Prevention and Institute of Social and Preventive Medicine of the University of Basel [¶]Immobarometer (n=1050), Neue Zürcher Zeitung [**]Swiss Health Survey (n=13'004), Swiss Federal Office of Statistics
n.a.= data not available yet, survey is underway

100

www.ingramcontent.com/pod-product-compliance
Lightning Source LLC
Chambersburg PA
CBHW021118210326
41598CB00017B/1494